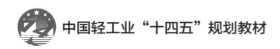
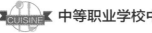

烹饪英语
（第二版）

PENGREN YINGYU
(DI ER BAN)

宋洁◎主编

图书在版编目（CIP）数据

烹饪英语/宋洁主编. —2版. —北京：中国轻工业出版社，2023.7
中等职业学校中餐烹饪专业教材
ISBN 978-7-5184-2931-8

Ⅰ.①烹… Ⅱ.①宋… Ⅲ.①烹饪—英语—中等专业学校—教材 Ⅳ.①TS972.1

中国版本图书馆CIP数据核字（2020）第041126号

责任编辑：史祖福　贺晓琴　　责任终审：张乃柬　　整体设计：锋尚设计
策划编辑：史祖福　　　　　　责任校对：吴大朋　　责任监印：张京华

出版发行：中国轻工业出版社（北京东长安街6号，邮编：100740）
印　　刷：三河市国英印务有限公司
经　　销：各地新华书店
版　　次：2023年7月第2版第3次印刷
开　　本：787×1092　1/16　印张：13
字　　数：291千字
书　　号：ISBN 978-7-5184-2931-8　定价：39.80元
邮购电话：010-65241695
发行电话：010-85119835　传真：85113293
网　　址：http://www.chlip.com.cn
Email：club@chlip.com.cn
如发现图书残缺请与我社邮购联系调换
231070J3C203ZBW

PREFACE 第二版前言

《烹饪英语》出版转眼八年多了，八年里，《烹饪英语》一书作为烹饪专业英语教材和选修课教材应用于全国众多课堂上。八年里，也收到了许多一线教师使用该教材的建议和心得，不胜感激。结合八年里本教材的使用情况和使用教师的意见，同时考虑烹饪行业的发展需求，现将该教材进行相应的调整。

本教材保留基础模块、职业模块和拓展模块三大模块。三大模块的功能不变，基础模块通过菜名公式为载体完成词汇积累，职业模块用各个烹饪相关主题对话完成对所学词汇的巩固及进一步应用，拓展模块以中餐烹饪英语知识点为基础向西餐烹饪英语知识点进一步拓展，包括从字词句到对话、从点到面的拓展。所以从基础模块到职业模块是按语言学习规律的纵向拓展，而拓展模块是对基础模块和职业模块内容的横向拓展。

另外，根据使用情况和行业发展，对各个模块内容做了如下调整。

（1）对基础模块中的词汇，特别是项目一中的词汇和菜名公式进行删减，保留基础和核心部分，其他内容作为拓展学习词汇，帮助学生更加有效地进行分层学习。

（2）同时，作为烹饪的主体——厨师，对岗位的分工、责任和安全等问题的了解和熟知属于基础职业素养，也是对食品安全的保证。因此，在基础模块中加入了厨师岗位英语项目，主要包括岗位设置、厨师职责和厨房安全的相关英语知识。

（3）在职业模块中，删除了西餐主题对话，并根据删除内容对相关习题进行调整，保证该模块重点是对基础模块的中餐烹饪英语的训练、巩固和提升。也因此在该模块中加入了厨师岗位英语的相关主题对话，对应基础模块中增设的厨师岗位英语部分。

（4）拓展模块的重点是对西餐英语相关知识点的学习，设置调整为从词汇到段落的学习过渡，与全书的设计思路统一。因此增设项目十一西餐主题口语对话，作为对项目十所学专业词汇的巩固提升。另外，由于近几年来烘焙行业发展迅猛，人们对烘焙产品的接受度越来越高，烘焙产品已经成为许多人日常饮食的一部分，因此在项目十中专门增设了烘焙英语的相关知识点，并在项目十一的菜肴制作过程的主题对话中进一步巩固。

本教材由宋洁担任主编，王晟兆、朱泓、周丹霞、王文媛担任副主编，郭遐曦参与本书编写，其中再版调整的内容由宋洁和王晟兆共同完成。

再次感谢中国轻工业出版社有限公司史祖福等编辑八年来对本教材的付出和帮助，及时提供教材使用者的反馈和修改意见，帮助教材更新改进，以符合时代发展和教学需求。同时感谢八年来广大专家、专业教师对本教材的支持及修改建议。此外，本次再版内容同样选录了许多图片资料，再次向原作者表示衷心的感谢。

由于编者水平有限,教材的进一步完善离不开广大专家和专业教师的宝贵意见,仍希望大家不吝赐教,我们编写团队会继续努力!

编　者

2020年4月

PREFACE 第一版前言

随着经济的迅速发展，人民的生活水平不断提高，中国的餐饮业呈现出一派蒸蒸日上、欣欣向荣的景象。餐饮业的发展进一步推动了烹饪专业职业教育的迅速壮大和发展。近年来，一方面，各相关学校的烹饪专业在校学生人数都呈增长态势，并且毕业生供不应求；另一方面，由于改革开放的进一步深入和各类城市的国际化程度日益提高，对烹饪等餐饮行业的从业人员提出了新的要求，既有烹饪专业技能，又懂烹饪英语的毕业生"一将难求"。在这种情况下，一些中等职业学校在烹饪专业中开始开设烹饪英语课程。本教材就在这样的背景下应运而生。

本教材以中式烹饪为重点进行讲解，共分三个模块，即基础模块、职业模块和拓展模块。

（1）基础模块　以菜名英文表述的常用公式为主线展开，引出烹饪专业相关的基础内容。在概述部分着重激发学生学习本课程的兴趣，并通过公式的引出让学生了解到学习后续章节的重要性，促使学生认真学习后续章节。最后通过小结，使各知识点之间联系得更为紧密并巩固所学知识。

（2）职业模块　主要由对话组成，旨在锻炼学生的口语表达能力，加强学生对外交流的能力。

（3）拓展模块　本模块对中国常见菜肴的英文表达进行了汇编，以便于学生学习菜名公式后进行实际应用时对照，完善了本教材的内容。

本教材每一小节的学习都会加入学习背景等相关知识链接，这样不仅可以让学生在学习时与本专业的专业知识联系起来，也可以帮助英语专业教师在教授烹饪英语课程时了解一些相关的烹饪专业知识。

本教材紧扣烹饪工作的实际需要，努力贴近学生生活，力求体现以下特点。

（1）职业性　教材的内容设计与学生学习的专业内容和未来的工作内容都是息息相关的。

（2）任务性　教材一开始就向学生展示了一个可以普遍使用的菜名公式，其他内容均是在此基础上展开的，学生一开始就明确了学习的任务，每一步的学习都有明确的目的性，而且能够让学生在以后的工作岗位中将所学的英语学以致用，这也是本教材编写的初衷，同时是学生学习的基本目的和本质任务。

（3）实用性　本教材注重与职业相关，与专业相关，致力于提高学生日后工作中的英语交际能力，有很强的实用性。

（4）趣味性　考虑到学生的英语基础和学习兴趣等问题，在教材的编写中，我们也更多地利用图片、任务教学、游戏设置等，增强教材的趣味性，以提高学生学习的兴趣和动力。

（5）易学性　由于中职学生的英语基础比较薄弱，在课程内容设计时，我们秉承"实用为主，够用为度"的原则，争取让学生用相对最基础的英语知识学到最多的专业内容。

（6）创新性　本教材遵循课程改革的原则和当前职业教学理论的要求，无论从教材总体大纲的设计还是具体内容的展开上，都对以往教材的形式有所创新和突破。

本教材由宋洁担任主编，朱泓、周丹霞、王文媛担任副主编，郭遐曦、王晟兆参加了本书的编写。

特别感谢金华商业学校副校长俞一夫和中国轻工业出版社的编辑在编写过程中给予的大力支持。此外，在本教材的编写过程中，参阅并选录了许多的图片资料，在此向原作者表示衷心的感谢。

由于编者水平有限，加之编写时间较为仓促，教材中难免会有疏漏和不足之处，希望相关专家、专业教师和广大读者不吝赐教。

编　者
2012年3月

CONTENTS 目录

Part I

第一部分 基础模块

项目一　中式菜肴菜名的英文表述 / 2
一、菜名英文表述的基本方法概述 / 2
二、常用烹饪方法的英文表述 / 7
三、主要烹饪原料的英文表述 / 17
四、常用烹饪调料的英文表述 / 35
五、常用菜肴风味的英文表述 / 40
六、菜名的英文表述小结 / 46

项目二　厨房常用设备与工具 / 50
一、刀具的英文表述 / 50
二、加热设备的英文表述 / 56
三、盛器的英文表述 / 62
四、辅助设备的英文表述 / 68

项目三　厨师岗位英语 / 73
一、厨房岗位的英文表述 / 74
二、厨师职责的英文表述 / 76
三、厨房安全与卫生的英文表述 / 78

Part II

第二部分 职业模块——烹饪英语常用对话

项目四　点餐 / 81
一、散餐；按菜单点菜 / 81
二、自助餐 / 84
三、酒水 / 85
四、项目四总练习 / 87

项目五　菜肴特色介绍 / 88
一、川菜 / 88
二、鲁菜 / 91
三、淮扬菜 / 94
四、项目五总练习 / 96

项目六　菜肴制作过程介绍 / 97
一、中餐 / 97
二、项目六总练习 / 102

项目七　厨房介绍 / 102
　一、中式厨房 / 102
　二、厨房设备与工具 / 104
　三、项目七总练习 / 110

项目八　面试英语（烹饪）/ 111
　一、自我介绍 / 111
　二、面试对话 / 113
　三、项目八总练习 / 115

项目九　厨师岗位 / 116
　一、从学徒工开始做起 / 116
　二、协作在厨房 / 118
　三、厨房准则 / 120
　四、厨房安全 / 122
　五、项目九总练习 / 123

第三部分　拓展模块

项目十　西式餐点简介 / 126
　一、西餐基本烹饪法 / 126
　二、西餐食品原料 / 129
　三、西餐烘焙简介 / 133
　四、西餐用餐礼仪 / 138

项目十一　西餐主题口语对话 / 143
　一、点餐 / 143
　二、菜肴制作 / 146
　三、项目十一总练习 / 151

第四部分　附录

　一、常见中式菜肴英文表述 / 153
　二、常见西式菜肴英文表述 / 160
　三、课文相关练习答案 / 163
　四、单词表 / 191

参考文献 / 200

Part I Basic Module
第一部分　基础模块

◎ **项目一**
中式菜肴菜名的英文表述
The English Expressions of Chinese Dishes

◎ **项目二**
厨房常用设备与工具
Kitchen Utensil and Cooking Equipment

◎ **项目三**
厨师岗位英语
English for the Cook on the Post

项目一 中式菜肴菜名的英文表述
The English Expressions of Chinese Dishes

一、菜名英文表述的基本方法概述
The Outline of the English Expressions of Chinese Dishes

教学指引 Teaching Guideline

中国菜肴五花八门且数量众多，其中往往有着悠久的历史，并且蕴含着深厚的文化底蕴。那么，如何使学生很好地理解中餐菜肴的内涵，同时如何更好地去记忆以及翻译中餐菜肴的名称，是我们需要解决的问题。中餐菜肴的记忆与翻译有一定的规律可循，本项目的主要内容就是对目前较为流行的中餐菜肴英文名称的构成规则与翻译规则进行一个归纳总结，让学生了解菜肴名称规律后再对其规律构成的公式进行具体学习。

知识背景介绍 The Introduction of Knowledge Background

中餐菜肴种类繁杂，但命名方式大致可以分为以下六类，通过不同的分类方法，可以采用不同的方式进行记忆与翻译。

（1）以菜肴的原料命名　此类菜名仅仅包括了菜肴的原料，但有的兼顾菜肴的风味或形状。例如，杭椒牛柳、龙井虾仁、香辣蟹、鱼香肉丝等。

（2）以菜肴的烹饪技法与原料命名　此类菜名包括了菜肴的烹调手段与原料。例如，软炸里脊、回锅肉、油焖大虾等。

（3）以菜肴的烹饪技法、形状与原料命名　此类菜名不仅包括了菜肴的烹调手段与原料，而且包含了菜肴的形状。例如，清炒墨鱼花、葱烧鱼片、XO酱炒牛柳等。

（4）以菜肴的烹饪技法与数字命名　此类菜名因菜肴主料较多，所以仅包括了菜肴的烹调手段与原料的大约数量。例如，扒四宝、糟熘三白、卤十件等。

（5）以菜肴扬名的地域或菜肴创始人命名　以此方法命名的菜肴多数为名菜。例如，北京烤鸭、扬州炒饭、东坡肘子、毛氏红烧肉等。

（6）以历史典故、民间传说与含有文化韵味的词语命名　此类菜名大多较为笼统，完全以美好的寓意与诗意来命名,部分含有菜肴的形状、原料与烹饪技法。例如,全家福、龙凤呈祥、佛跳墙、宋嫂鱼羹、叫花鸡等。

学生学习 Student Learning

导入 Lead-in

(1)结合图片找出下列菜肴的主料、辅料、汤汁分别是什么,填入表格。

玉米肉丸

姜汁鲜鱿

美味牛筋

麻辣肚丝

皮蛋豆腐

	主料	辅料	汤汁
玉米肉丸			
姜汁鲜鱿			
美味牛筋			
麻辣肚丝			
皮蛋豆腐			

(2)根据给出菜肴的英文译名,试着推断菜肴英文翻译有何规律。

玉米肉丸:Meatballs(肉丸)and Corn(玉米)

姜汁鲜鱿:Fresh(新鲜的)Squid(鱿鱼)in Ginger(生姜)Sauce(汤汁)

美味牛筋:Beef(牛肉)Tendon(筋腱)

麻辣肚丝:Shredded(切丝的)Pig Tripe(肚子)in Chili(辣椒)Sauce(汤汁)

皮蛋豆腐:Tofu(豆腐)with Preserved(腌制的)Eggs(蛋)

基础公式学习 Basic Expressions Learning

根据"导入"所做的练习，我们可以得出一个菜肴菜名表达的公式：主料+形状+介词（and/with）+辅料+介词（in/with）+汤汁。

我们可以再来看几个例子。

白灵菇扣鸭掌：Mushrooms with Duck Feet

解析：$\dfrac{\text{Mushrooms（主料）}+\text{with（介词）}+\text{Duck Feet（辅料）}}{\text{蘑菇}\qquad\qquad\qquad\qquad\qquad\text{鸭掌}}$

葱油鹅肝：Goose Liver with Scallion in Chili Oil

解析：$\dfrac{\text{Goose Liver（主料）}+\text{with（介词）}+\text{Scallion（辅料）}+\text{in（介词）}+\text{Chili Oil（汤汁）}}{\text{鹅}\quad\text{肝脏}\qquad\qquad\qquad\qquad\text{葱}\qquad\qquad\qquad\qquad\text{辣椒油}}$

更多例子
扫描二维码获取

◎ 细解 Explanation

① 这个基础公式最主要的成分是"主料"，其余成分均有被一起省去或部分省略的可能。

② 表示菜肴"形状"的词主要如下。

cube [kju:b]	n.（立方）块
chunk [tʃʌŋk]	n.（较大的，不规则的）块
roll [rəul]	n. 卷
filet [fi'lei]	n. 肉片；鱼片
slice [slais]	n. 薄片

◎ 思考题 Questions

结合课本给出的例子思考以下问题。

① 主料与辅料之间的介词一般用and或者with，到底什么时候用and，什么时候用with呢？

② 汤汁前面的介词一般用in或者with，分别又是在什么情况下使用呢？

◎ 细解 Explanation

① 课文例子中主料与辅料之间全部用了with，现实表达中也是这样，主料和辅料之间的介词基本上都是with，用and的只占了很小一部分，且多用在汤、煲一类的菜肴中。大家可以这样理解，当辅料在菜肴中的作用或比重与主料差不多的时候，或者是主料与辅料难以区别的时候，我们才用and，其余情况基本用with。

例如，竹笋青豆：$\dfrac{\text{Bamboo Shoots \textit{and} Green Beans}}{\text{竹笋}\qquad\qquad\qquad\text{青豆}}$

翠豆玉米粒：$\dfrac{\text{Sautéed Green Peas \textit{and} Sweet Corn}}{\text{油煎的}\qquad\text{豌豆}\qquad\qquad\text{甜玉米}}$

更多例子
扫描二维码获取

② 汤汁前的连词使用一般有这样的规律：
在菜名中，如果主料是浸泡在汤汁中的，使用in连接。

例如，酒醉排骨：$\underline{\text{Spare Ribs}}\ \underline{\textit{in}}\ \underline{\text{Wine Sauce}}$
　　　　　　　　排骨　　　酒　汤汁

香橙炖官燕：$\underline{\text{Braised Bird's Nest}}\ \underline{\textit{in}}\ \underline{\text{Orange Sauce}}$
　　　　　　炖、焖　燕窝　　橙子　汤汁

酒醉排骨

在菜名中，如果汤汁或者蘸料与主料是分开的，或者是后浇于主菜上，则一般用with连接。

例如，香辣猪扒：$\underline{\text{Grilled Pork}}\ \underline{\textit{with}}\ \underline{\text{Spicy Sauce}}$
　　　　　　　烤、炙　猪肉　　香辣的　汤汁

茄汁鱼片：$\underline{\text{Fish Filets}}\ \underline{\textit{with}}\ \underline{\text{Tomato Sauce}}$
　　　　　　鱼　片　　　番茄　汤汁

茄汁鱼片

📚 提示 Tips

上述公式是菜肴表达的基本公式，也是本项目学习的基础公式，还有很多菜肴是通过在这个基础公式上衍生出来的其他公式加以翻译的，所以记住这个基础公式，学会一些常用主料、辅料和汤汁的英文名称，就可以说出很多菜肴的英文名了。详细表达我们在后面的课文中马上会学到。

◎ 巩固练习 Exercise to Consolidate

小辉点了以下三道菜，他发现，所点的三个菜的英文名分别用到了老师今天讲到的三个介词in、and、with。你能根据所学知识判断以下每一幅图对应的分别是哪个介词吗？请将该介词填入图片下方的括号内。（What preposition can be used?）

（　　）

（　　）

（　　）

需要掌握的字词 Words and Expressions You Need to Grasp

beef [bi:f] *n.* 牛肉

pig [pig] *n.* 猪；猪肉

pork [pɔ:k] *n.* 猪肉

tofu ['təufu:] *n.* 豆腐

ginger ['dʒindʒə] *n.* 生姜

chili ['tʃili] *n.* 红辣椒，辣椒

chili sauce 辣酱油；辣味番茄酱

sauce [sɔ:s] *n.* 酱油；少司；调味汁

shredded [ʃredid] *adj.* 切丝的

提示 Tips

大家都知道pig是"猪"的意思，但也要记得它还有"猪肉"的意思。pork与pig的具体区别，我们会在烹饪原料中详细阐述。

美食欣赏 Enjoy the Delicacies

豆豉鲫鱼　　　　香辣茄子

桂花山药　　　　酸辣瓜条　　　　姜汁蜇皮

练习 Exercises

根据所给所学和上图下文提示，写出以下菜肴的英文名称。（Translate the Chinese dishes into English according to what you have learnt, the pictures above and the clues given below.）

Crucian Carp（鲫鱼）　　　　Eggplant（茄子）

Cucumber（黄瓜）　　　　Hot and Sour Sauce（酸辣汁）

Chinese Yam（山药）　　　　　　Jellyfish（海蜇）

Osmanthus（桂花）　　　　　　　Chili Sauce（辣酱油）

Black Bean（豆豉、黑豆）

桂花山药_____

豆豉鲫鱼_____

酸辣瓜条_____

香辣茄子_____

姜汁蜇皮_____

二、常用烹饪方法的英文表述
The English Expressions of Common Chinese Cookery

教学指引 Teaching Guideline

中餐的烹饪方法五花八门，近年来，因为吸收了部分西餐的烹饪方法而变得更加丰富多彩，因为在以后学习的菜名拓展公式中涉及烹饪方法，所以本项目的学习目的是，通过介绍几种主要中餐烹饪方法的英文表述，使学生对上一项目所学习的中餐菜肴英文名称的构成规则与翻译规则加深理解与记忆，并且可以举一反三。由于本节学习的烹饪技法内容比较多，建议任课教师将烹饪技法和公式穿插学习，这样有利于学生的理解和记忆。在课时分配上，也可以根据学生的具体情况，灵活分配。

知识背景介绍 The Introduction of Knowledge Background

烹饪方法的种类如下。

按菜品温度 { 热菜技法： 烧、扒、煨、炖、烩、焖、汆、煮、炒、炸、爆、煎、贴、烹、蒸、烤、拔丝、蜜汁、挂霜等

凉菜技法： 拌、腌、炝、冻、酱、卤等 }

按传热介质 {
　水烹法：　烧、扒、煨、炖、烩、焖、汆、煮、蜜汁等
　油烹法：　炒、炸、爆、煎、贴、烹、拔丝、挂霜等
　汽烹法：　蒸和熏等
　辐射法：　烤和微波烹调等
　其　他：　盐和石烹等
}

学生学习 Student Learning

导入 Lead-in

（1）合上书本，比一比谁能说出更多的烹饪方法。
（2）讨论一下，哪些是烹饪中常用的方法。

常用烹饪方法的英文表达 The English Expressions of Common Chinese Cookery

由于中西方的语言、文化、烹饪技法都存在差异，中式烹调中的方法与英文并不能一一对应，有的甚至没有与之相对应的英文词汇，只能找相近词语代替。因此，本节会把英文单词的本意及它可以代表中文中的哪些烹饪技法做一个说明。

（1）sauté [ˈsəutei]

英文本意：炒；嫩煎

可代表的中式烹饪技法：炒；炝；爆；煸；熘

延伸：sauté 一般是指用少许热油快速地煎炒或者用油将菜肴烧至棕色。

（2）scramble [ˈskræmbl]

英文本意：炒（鸡蛋）、煎（鸡蛋）

可代表的中式烹饪技法：炒、煎（鸡蛋）

延伸：这个动词后面的烹饪对象一般都是鸡蛋。

（3）fry [frai]

英文本意：油炸；油煎

可代表的中式烹饪技法：炸；煎；贴；烹

延伸：fry 一般是用热油炸，特别是指在专门的容器中用一定量的油炸。以fry为中心词，可具体衍生出如下各种烹饪技法。

注解①：pan-fry中的pan是"平底锅"的意思，所以这里的煎一般是指用平底锅煎。

（4）simmer ['simə]

英文本意：炖、煨

可代表的中式烹饪技法：炖、煨

延伸：simmer指用低于沸点的温度烹制液体食物或者是带有液体的食物，慢慢加热，常常需要很长时间。

（5）stew [stju:, stu:]

英文本意：煲、炖、焖

可代表的中式烹饪技法：煲、炖、焖

延伸：stew指把食物放在汤、汁中用低温慢慢加热，这点与simmer基本同义，但stew常指把鱼、肉和蔬菜在肉汁或调料中混合在一起做出来的食物，也就是说用stew这种烹饪方法的菜肴通常用的原料相对比较多一些，在时间上也会比simmer相对更长一些。

$$\begin{cases} \text{stewing in water：将食物放在水中煲} \\ \text{stewing out of water：隔水炖，用于炖补品} \\ \text{stewing in gravy}[2]\text{：放进调味汁里煮，就是"卤"了} \end{cases}$$

注解②：gravy ['greivi] *n.* 肉汁，肉卤，调味汁。

（6）braise [breiz]

英文本意：（用文火）炖；烧

可代表的中式烹饪技法：炖；焖；烩；烧；浸；扒

延伸：braise指用少量的油把肉（通常还混有蔬菜）放在有盖的密闭容器中，再加入少量汤煨、炖的烹饪方法。有时候我们还可以将它的烹饪过程理解为：sauté+simmer。

（7）scald [skɔ:ld]

英文本意：煮沸；烫

可代表的中式烹饪技法：白灼；烫

（8）boil [bɔil]

英文本意：煮沸；在沸水中煮

可代表的中式烹饪技法：煮

延伸：煮的特点是使食物保持鲜嫩，煮一般还可以分为以下两种。

$$\begin{cases} \text{instant}[3]\text{-boiling：即煮即食。比如北方的"火锅"，广东的"打边炉"} \\ \text{quick}[3]\text{-boiling：快煮。将煮沸的汤浇入盛有食物的器皿里，或将食物投入煮沸了的水中，然后慢火煮热} \end{cases}$$

注解③：instant ['instənt] *adj.* 立即的，即刻的；quick [kwik] *adj.* 快的，迅速的。

（9）marinate ['mærineit]

英文本意：浸泡在卤汁中，在食物上浇上卤汁

可代表的中式烹饪技法：酱

（10）ferment ['fə:ment]

英文本意：发酵

可代表的中式烹饪技法：酿

（11）pickle ['pikl]

英文本意：（用盐水或者醋汁）腌制水果或者蔬菜

可代表的中式烹饪技法：泡，腌，盐卤

（12）steam [sti:m]

英文本意：蒸

可代表的中式烹饪技法：蒸

（13）smoke [sməuk]

英文本意：熏

可代表的中式烹饪技法：熏

（14）有关于烤的词

{
bake [beik] 烤，烘焙，是在密闭的烘具中烘，不用明火，一般不外加油或汤汁

grill[4] [gril] 烧，烤

broil [brɔil] 烤，炙
} 是与明火直接接触或放在铁架等烧热的炊具表面上烤

roast [rəust] 烤，烘，不用明火，通常外加油或汤汁
}

注解④：grill与broil同义，只不过broil是美式英语的说法，但在中文菜名中，用grill的占大多数。

◎ 巩固练习 Exercise to Consolidate

① 连一连（Match Task）

请帮忙将逃跑的气球固定在正确的位置上。

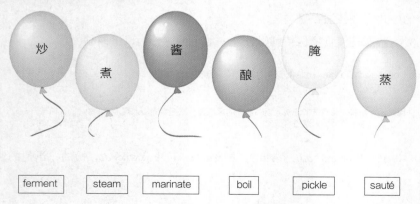

② 以下单词中，哪几个单词与下图厨房用具有关，请将相关的单词放入锅中方框处。（Put the right words into the textbooks.）

roast
simmer
stew
smoke
braise
scald
bake

提示 Tips

① 大家尽可能地去理解并记忆这些烹饪技法单词的英文本意，因为只有掌握英语单词的本意，翻译时才能做到准确，才能使翻译出的英文被理解，不产生歧义。

比如"酱豆腐"，就是我们平时说的"豆腐乳"，它的烹饪方法是"酱"，如果照搬中文原词，应该用"marinate"，但是豆腐乳的制作有一个发酵的过程，所以应该用"ferment"更为贴切。故翻译为"Fermented Bean Curd"。"葱烤银鳕鱼"中"烤"的英文单词有bake, grill, broil, roast，但是这道菜中的"烤"是指放在烧热的炊具上烤，所以这道菜名可翻译为"Grilled Codfish with Scallion"。因此切忌望文生义、生搬硬套。

② 因为有些菜肴的烹调工序比较复杂和繁琐，在为其选择烹饪技法单词时，主要还是选择工序中最主要或者对菜肴成形最重要的烹饪技法的英文单词。

公式学习 Expressions Learning

◎ 思考题 Questions

结合上文提示中的内容，思考如下两个问题。

① 运用烹饪方法写出的菜名中，对应烹饪技法的英文单词发生了什么变化，位置上又有什么规律（先不要去管其他菜名等生词，就看看关于烹饪技法的单词）。

② 这些菜名的表达跟我们刚学习的基础公式又有什么联系？

酱豆腐：$\dfrac{\text{Fermented Bean Curd}}{\text{豆腐}}$

更多例子
扫描二维码获取

葱烤银鳕鱼：Grilled Codfish with Scallion
　　　　　　　　　鳕鱼　　　葱

香辣虾：Fried Shrimps in Hot Spicy Sauce
　　　　　基围虾　　　香辣汁

基础公式：主料+形状+介词（and/with）+辅料+介词（in/with）+汤汁

葱烤银鳕鱼

◎ 拓展公式学习 Extended Expressions Learning

通过对思考题的研究，我们可以根据烹饪技法得出如下拓展公式。

拓展公式一：烹饪技法（过去分词）+基础公式

公式中的烹饪技法所用的单词要用过去分词的形式，只有roast大多数情况用原形，所以这个公式还可以进一步表示为：烹饪技法的过去分词+基础公式。我们可以多看几个例子。

焗肉排：Baked Spare Ribs

解析：Baked（烹饪技法的过去分词）+ Spare Ribs（主料）
　　　　　　　　　　　　　　　　　　　排骨

煎猪柳

煎猪柳：Pan-Fried Pork Filet

解析：Pan-Fried（烹饪技法的过去分词）+ Pork（主料）+ Filet（形状）
　　　　　　　　　　　　　　　　　　　猪肉　　　　　肉片

更多例子
扫描二维码获取

番茄炖牛腩：Braised Beef Brisket with Tomato

解析：Braised（烹饪技法的过去分词）+ Beef Brisket（主料）+ with（介词）+ Tomato（辅料）
　　　　　　　　　　　　　　　　　　　牛腩　　　　　　　　　　　　　　番茄

豉汁牛仔骨：Steamed Beef Ribs in Black Bean Sauce

解析：

Steamed（烹饪技法的过去分词）+Beef Ribs（主料）+in（介词）+Black Bean Sauce（汤汁）
　　　　　　　　　　　　　　　　牛排骨　　　　　　　　　　　　豆豉汁
　　　　　　　　　　　　　　　　　　　　基本公式

◎ 需要掌握的字词 Words and Expressions You Need to Grasp

sauté [ˈsəutei] v. 炒；嫩煎　　　　　fry [frai] v. 油炸；油煎

烹饪技法表格
扫描二维码获取

scald [skɔ:ld] v. 煮沸；烫
grill [gril] v. 烧，烤
bake [beik] v. 烤，烘焙
simmer ['simə] v. 炖、煨
braise [breiz] v.（用文火）炖；烧
steam [sti:m] v. 蒸
marinate ['mærineit] v. 浸泡在卤汁中；在鱼或肉上浇上卤汁
pickle ['pikl] v.（用盐水或者醋汁）腌制水果或者蔬菜

boil [bɔil] v. 煮沸；在沸水中煮
broil [brɔil] v. 烤，炙
roast [rəust] v. 烤，烘
stew [stju:, stu:] v. 煲，炖，焖
ferment ['fə:ment] v. 发酵
smoke [sməuk] v. 熏

美食欣赏 Enjoy the Delicacies

火爆川椒鸭舌

卤水大肠

当红炸子鸡

白灼西蓝花

萝卜干腊肉

浓汤娃娃菜

香辣虾

练习 Exercises

结合以上图片，分组讨论以下菜肴主要运用了什么烹饪技法，用所学英文单词表示。（Discuss in groups and try to judge what kind of cooking skill is mainly used in the dishes below according to the pictures above, and then name the cooking skill in English.）

卤水大肠　_____　　　　浓汤娃娃菜　_____

烟熏蜜汁肋排	_____	清蒸火腿鸡片	_____
酸黄瓜条	_____	烤银丝卷	_____
干豆角回锅肉	_____	烤全羊	_____
萝卜干腊肉	_____	白灼西蓝花	_____
当红炸子鸡	_____	香辣虾	_____
孜然烤牛肉	_____	火爆川椒鸭舌	_____
鲍鱼红烧肉	_____	炸肉茄盒	_____
药味炖生中虾	_____	烫青菜	_____
冬菜扣肉	_____	鲜奶煮蛋	_____

更多练习
扫描二维码获取

拓展学习 Extended Learning

我们回顾一下学习的拓展公式一：

烹饪技法（过去分词）+基础公式=烹饪技法（过去分词）+主料+形状+介词（and/with）+辅料+介词（in/with）+汤汁

其实从英语词法的角度来解释，表示烹饪技法的动词变成过去分词后的作用是修饰后面的主料，说明主料的烹调方法，也就是说，拓展公式中烹饪技法的过去分词的作用是修饰，相当于一个形容词的作用。

由此我们可以得出第二个拓展公式。

拓展公式二：修饰词+基础公式

◎ **细解 Explanation**

拓展公式中"修饰词"可以用的范围很广，除了表示烹饪技法的修饰词外，还有如下几种。

① 用修饰词表示刀法

例如，青豌豆肉丁：Sautéed Diced Pork with Green Peas

解析：Sautéed（修饰词）+ Diced（修饰词）+ Pork（主料）+ with（介词）+ Green Peas（辅料）
　　　　　　　　　　　　切粒的　　　　　猪肉　　　　　　　　　　　青豌豆
　　　　　　　　　　　　　　　　　　　　　基础公式

表示刀法的常用修饰词如下。

diced [daɪst] *adj.* 切粒的
mashed [mæʃt] *adj.* 捣碎的；捣烂的
minced [mɪnst] *adj.* 切碎的；切成末的
shredded [ʃredɪd] *adj.* 切丝的；切碎的
sliced [slaɪst] *adj.* 切成薄片的

② 用修饰词表示口感或口味

例如，脆皮鸡：Crispy Chicken

解析：$\dfrac{\text{Crispy（修饰词）} + \text{Chicken（主料）}}{\text{酥脆的（表口感）} \quad \text{鸡肉}}$
基础公式

脆皮鸡

香辣芙蓉鱼：Spicy Fish Filets with Egg White

解析：$\dfrac{\text{Spicy（修饰词）} + \text{Fish（主料）} + \text{Filets（形状）} + \text{with（连词）} + \text{Egg White（辅料）}}{\text{辛辣的（表口味）} \quad \text{鱼} \quad \text{片} \quad \text{蛋白}}$
基础公式

表示口感和口味的常用词如下。

| bitter [ˈbitə] adj. 苦的 |
| crisp [krisp] adj. 脆的 |
| hot [hɔt] adj. 辣的 |
| sour [ˈsauə] adj. 酸的 |
| spicy [ˈspaisi] adj. 辛辣的；香的 |
| sweet [swi:t] adj. 甜的 |
| tender [ˈtendə] adj. 嫩的 |

③ 用人名来修饰，体现菜肴特色

例如，东坡肉：Dongpo's Braised Pork

解析：$\dfrac{\text{Dongpo's（修饰词）} + \text{Braised（修饰词）} + \text{Pork（主料）}}{\text{东坡（人名）} \quad \text{猪肉}}$
基础公式

东坡肉

左公鸡：General Tso's Chicken

解析：$\dfrac{\text{General Tso's（修饰词）} + \text{Chicken（主料）}}{\text{将军左宗棠} \quad \text{鸡肉}}$
基础公式

④ 介绍一个凉拌的烹饪技法

例如，拌双耳：Tossed Black and White Fungus

左公鸡

解析：<u>Tossed（修饰词）</u> + <u>Black and White Fungus（主料）</u>
　　　凉拌的　　　　　　　黑木耳和白木耳
　　　　　　　　　　　　　　基础公式

注解：toss [tɔ:s] *v*. 拌。

拌双耳

> ## 提示 Tips
>
> ① 在拓展公式二中提到的各种类型的修饰词可以单独使用，也可以一起使用。例如，青豌豆肉丁：<u>Sautéed Diced</u> Pork with Green Peas，就是表示烹饪技法和刀法的修饰词一起使用的例子。
>
> ② 拓展公式一其实可以看成是拓展公式二的子公式，由于这个子公式运用的频率最高，才将其单独列出。

◎ 需要掌握的字词 Words and Expressions You Need to Grasp

bitter ['bitə] *adj*. 苦的
diced [daist] *adj*. 切粒的
minced [minst] *adj*. 切碎的；切成末的
shredded [ʃredid] *adj*. 切丝的；切碎的
sour ['sauə] *adj*. 酸的
sweet [swi:t] *adj*. 甜的

crisp [krisp] *adj*. 脆的
hot [hɔt] *adj*. 辣的
mashed [mæʃt] *adj*. 捣碎的；捣烂的
sliced [slaist] *adj*. 切成薄片的
spicy ['spaisi] *adj*. 辛辣的；香的
tender ['tendə] *adj*. 嫩的

练习 Exercises

根据所给图片和提示，补充完整以下菜名。（Fill in the blanks according to the pictures and clues given.）

> ## 提示 Tips
>
> Sago 西米　　　　Coconut Milk 椰奶　　Beef 牛肉
> Potato 马铃薯　　Pork 猪肉　　　　　　Puff 泡芙
> Prawn 明虾，对虾　Duck 鸭肉　　　　　Soy Sauce 酱油
> Pepper 胡椒　　　Chili 辣椒　　　　　Chairman 主席

椰汁鲜果西米露　　　　　　五香牛肉　　　　　　　薯泥
_____Sago with Fresh Fruit in　　_____Roast Beef　　　_____Potato
Coconut Milk

肉松松饼　　　　　　　糖醋咕噜虾球　　　　　　香酱爆鸭丝
_____Pork Puff　　　　　_____Prawn　　　　　_____Duck in Soy Sauce

回锅肉片　　　　　　　毛氏红烧肉
Sautéed_____Pork with　　Chairman_____Stewed
Pepper and Chili　　　　Pork with Soy Sauce

三、主要烹饪原料的英文表述
The English Expressions of Cooking Material

教学指引 Teaching Guideline

　　烹饪原料学是烹饪专业、食品专业学生的重要专业基础课，是从事烹饪工作、临床营养研究等从业人员所必备的基础知识之一。因此烹饪英语中对烹饪原料英语的学习也是至关重要的。在菜肴表述的公式中，烹饪原料是最基础的元素之一，只有学会常用烹饪原料的英文表述，才有可能应用好公式。但是常用的烹饪原材料比较多，如果采用一股脑、填鸭式的教学方式，显然不符合教学原理，学生也难以学习。所以教师要合理安排好课时和授课内容，传授时与课文游戏及课外教师自身准备的材料结合起来，灵活教学，提高学生的学习兴趣和学习效率。

知识背景介绍 The Introduction of Knowledge Background

烹饪原料是指符合饮食要求、能满足人体营养需要，并通过烹饪手段制作各种食品的可食性食物原材料。常见的烹饪原料分类方法如下。

（1）按照烹饪原料在加工中的作用，分为主料、配料、调辅料。
（2）按照原料的来源分为动物性原料、植物性原料、矿物性原料、人工合成原料。
（3）按照原料的加工程度分为鲜活原料、干活原料、复制品原料。
（4）按照商品的体系分为粮食、蔬菜、果品、肉及肉制品、水产品、干货及干货制品、蛋奶及蛋奶制品、调味品等。

学生学习 Student Learning

烹饪原料——蔬菜类 Cooking Material—Vegetables
（1）豆类（Beans and Peas）

更多豆类
扫描二维码获取

beans	豆类；黄豆 [c]	[bi:n]
broad bean	蚕豆 [c]	broad [brɔ:d] *adj.* 宽的，辽阔的
French bean	四季豆；芸豆；扁豆 [c]	French [frentʃ] *adj.* 法国的，法国人的
green bean	青豆、四季豆 [c]	
soy bean	大豆 [c]	soy [sɔi] *n.* 大豆，豆酱
sour bean	酸豆角 [c]	sour ['sauə] *adj.* 酸的

续表

bean sprouts	豆芽 [pl]	sprout [spraut] n. 芽，苗芽
pea	豌豆 [c]	[pi:]
cowpea	豇豆 [c]	['kau,pi:]
snow pea	雪豆；荷兰豆 [c]	snow [snəu] n. 雪

提示 Tips

各个表格中有关名词的可数与不可数性质做一下区分。

（1）[c]：可数

（2）[u]：不可数，一般表格中不作任何注解的，默认该名词为不可数

（3）[pl]：以复数形式出现

（4）[u/c]：既可以可数也可以不可数，但不可数用法居多

相关豆类制品（Related Products）

tofu (bean curd)	豆腐	curd [kə:d] n. 凝乳状物
dried tofu	豆腐干	
dried cowpea	豇豆干 [c]	dried [draid] adj. 弄干了的
dried bean	干豆；干制豆类	
Douzhir (fermented bean drink)	豆汁儿	fermented [fə:'mentid] adj. 发酵的
preserved tofu	皮蛋	preserved [pri'zə:vd] adj. 腌制的
tofu skin	豆腐皮	skin [skin] n. 皮肤；外皮

更多豆类制品
扫描二维码获取

（2）瓜类；甜瓜（Melon）

cucumber	黄瓜 [u/c]		['kju:kʌmbə]
papaya	木瓜		[pə'pɑ:iə]
pumpkin	南瓜		['pʌmpkin]
white Gourd	冬瓜	white [wait] adj. 白色的	gourd [guəd] n. 葫芦
bitter Melon	苦瓜	bitter ['bitə] adj. 苦的	melon ['melən] n. 瓜，甜瓜

◎ 巩固练习 Exercise to Consolidate

国际部餐厅负责进货的是个英国小伙，你希望他帮助你进以下原料，请告诉他你需要原料的英文名称，并给出最新的市场价以防他买贵。（Find out the English name and the proper price.）

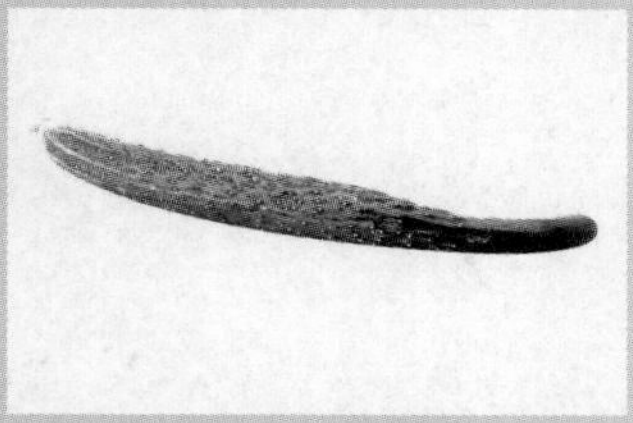

Name:＿＿＿＿＿＿＿＿＿＿＿＿＿＿＿＿

Price:＿＿＿＿＿＿＿＿＿＿＿＿＿＿＿＿

Name:＿＿＿＿＿＿＿＿＿＿＿＿＿＿＿＿

Price:＿＿＿＿＿＿＿＿＿＿＿＿＿＿＿＿

Name:＿＿＿＿＿＿＿＿＿＿＿＿＿＿＿＿

Price:＿＿＿＿＿＿＿＿＿＿＿＿＿＿＿＿

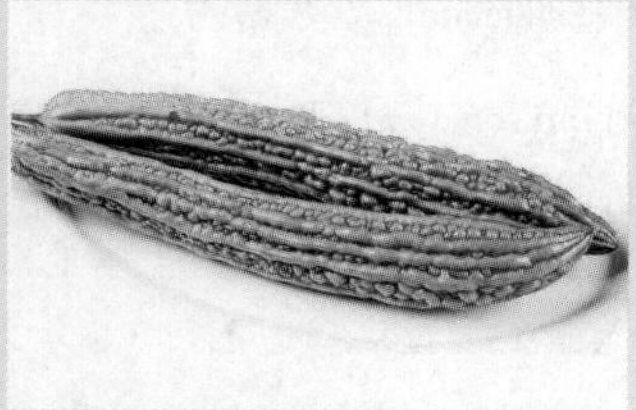

Name:＿＿＿＿＿＿＿＿＿＿＿＿＿＿＿＿

Price:＿＿＿＿＿＿＿＿＿＿＿＿＿＿＿＿

Name:＿＿＿＿＿＿＿＿＿＿＿＿＿＿＿＿

Price:＿＿＿＿＿＿＿＿＿＿＿＿＿＿＿＿

（3）绿叶蔬菜（Green Vegetables）

broccoli	西蓝花；花椰菜	['brɔkəli]
green vegetable	青菜	
celery	芹菜	['seləri]
cabbage	卷心菜	['kæbidʒ]
leek	韭葱 [c]	[li:k]
lettuce	莴苣；生菜 [u/c]	['letis]
spinach	菠菜	['spinidʒ]
crown daisy	茼蒿	crown [kraun] n. 花冠；顶点 daisy ['deizi] n. 菊科植物
Chinese toon	香椿	toon [tu:n] n. 红椿木

更多绿叶蔬菜
扫描二维码获取

◎ 巩固练习 Exercise to Consolidate

Name:_____

Price:_____

Name:_____

Price:_____

Name:_____

Price:_____

Name:_____

Price:_____

Name:_____

Price:_____

Name:_____

Price:_____

续表

Name:_____

Price:_____

Name:_____

Price:_____

Name:_____

Price:_____

Name:_____

Price:_____

（4）菌菇类（Mushrooms）

mushroom	蘑菇 [c]	[ˈmʌʃruːm]
black mushroom	冬菇、香菇 [c]	black [blæk] *adj.* 黑色的
straw mushroom	草菇 [c]	straw [strɔː] *n.* 稻草
tea tree mushroom	茶树菇 [c]	tea [tiː] tree [triː] *n.* 茶树

更多菌菇类
扫描二维码获取

◎ 巩固练习 Exercise to Consolidate

Name:_____

Price:_____

Name:_____

Price:_____

续表

	Name:_____
	Price:_____
	Name:_____
	Price:_____
	Name:_____
	Price:_____

（5）蔬菜腌制品（Vegetable Pickles）

更多蔬菜腌制品
扫描二维码获取

kimchi	朝鲜泡菜	[ˈkimtʃi]
pickled vegetable	榨菜，泡菜 [u/c]	pickled [ˈpikld] *adj.* 腌制的

（6）其他（Others）

eggplant	茄子	[ˈegplɑ:nt]
tomato	番茄 [u/c]	[təˈmɑ:təu]
onion	洋葱 [u/c]	[ˈʌnjən]

续表

potato	马铃薯 [u/c]	[pə'teitəu]
taro	芋头	['tɑːrəu]
truffle	松露 [u/c]	['trʌfl]
carrot	胡萝卜 [u/c]	['kærət]
ginger	姜	['dʒindʒə]
garlic	大蒜，蒜头	['gɑːlik]
scallion	葱	['skæljən]
Chinese yam	山药	yam [jæm] *n.* 甘薯
daylily (day lily)	黄花菜	['deilili]

更多例子
扫描二维码获取

美食欣赏 Enjoy the Delicacies

百合炒南瓜

脆皮豆腐

冬菇扒菜心

雪菜炒豆皮

◎ 巩固练习 Exercise to Consolidate

根据前面所学，翻译菜名。（Translate the Chinese dishes into English according to what you have learnt and the pictures above.）

百合炒南瓜_____ 炒芥蓝_____
翠豆玉米粒_____ 冬菇扒菜心_____
火腿炒蚕豆_____ 木瓜炖百合_____
清煎番茄_____ 番茄炒蛋_____
烫青菜_____ 青叶豆腐_____
白灼西蓝花_____ 脆皮豆腐_____
牛肝菌红烧豆腐_____ 雪菜炒豆皮_____
子姜鸭_____ 清炒/蒜蓉西蓝花_____

烹饪原料——肉类 Cooking Material—Meat

（1）烹饪中常用肉类（Meat in Common Use）

chicken ['tʃikin] 鸡肉　　　　duck [dʌk] 鸭肉

pig [pig] 猪肉　pork [pɔːk] 猪肉　beef [biːf] 牛肉

更多肉类
扫描二维码获取

提示 Tips

在英文中，同样指"猪肉"的意思，pig比pork涵盖的范围更广，pig可以指熏肉（bacon）、火腿肉（ham）和新鲜的猪肉（pork）。不过在中文菜肴的表达中，猪肉还是pork使用得更多。当表示是猪身体某个部分时（非内脏部分），大部分情况下用pig，比如：

腊肉炒香芹：Sautéed Preserved Pork with Celery

什烩肉：Roast Pork with Mixed Vegetables

咖喱肉：Curry Pork

腐乳猪蹄：Stewed Pig Feet with Preserved Tofu

芸豆焖猪尾：Braised Pigtails with French Beans

腐乳猪蹄

◎ 巩固练习 Exercise to Consolidate

一个粗心的服务生把下面家禽的肉的英文名称标签贴错了，你能将这些打乱的标签重新贴到对应的动物上吗？（Match the labels with the right pictures.）（填写到空格处）

chicken　　　goose　　　duck　　　beef　　　pork　　　mutton

（2）其他肉类（Other Meat）

venison ['venizən] 鹿肉	bullfrog ['bulfrɔg] 牛蛙
rabbit ['ræbit] 兔肉；兔子 [c]	sparrow ['spærəu] 麻雀 [c]

（3）动物内脏（Entrails）

heart [hɑ:t] 心 [c]	lung [lʌŋ] 肺 [c]
liver ['livə]（供食用的）肝	gizzard ['gizəd]（鸟等的）砂囊、胗 [c]
kidney ['kidni]（动物等可食用的）腰子 [u/c]	tripe [traip] 肚
intestine [in'testin] 肠 [c]	

◎ 例子 Example

chicken gizzard 鸡胗	pork lung 猪肺
pork intestine 猪大肠	beef tripe 牛肚

◎ 思考题 Questions

结合上面的例子想一想，如何用英文表达动物内脏？

◎ 细解 Explanation

一般用英文来表达动物内脏都以"动物种类+内脏的形式"为公式，就是上文第一点内容加第三点内容的形式，即"（1）+（3）"的形式或者也可以是"（2）+（3）"的形式，比如，猪肚：pork tripe；鸭胗：duck gizzard；鹅肝：goose liver。

（4）其他部位（Others）

brain [brein] 脑 [c]	
breast [brest] 胸部 [c]	
belly ['beli] 腹部的肉	rib [rib] 排骨；肋骨 [c]
wing [wiŋ] 翅膀 [c]	leg [leg]（猪、羊等）供食用的腿 [u/c]
hock [hɔk] 肘关节	
feet [fi:t] 脚 [pl]	ear [iə] 耳朵 [c]
tail [teil] 尾巴 [c]	blood curd [blʌd] [kə:d] 血旺

更多例子
扫描二维码获取

◎ 细解 Explanation

一般用英文来表达动物其他部位也是以"动物种类+部位"的形式。就是上文第一点内容加第四点内容的形式，即"（1）+（4）"的形式或者也可以说是"（2）+（4）"的形式，比如，鸡翅：chicken wings；牛舌：ox tongue；鸡胸：chicken breast；鹅掌：goose feet。

注：在本页附图中给出了鸡的各个部分，十分具体，有兴趣的同学可以了解一下，大家只要掌握上面列出的常用的词汇即可。

美食欣赏 Enjoy the Delicacies

 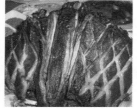  

番茄炖牛腩　　　　烤羊里脊　　　　鸿运蒸凤爪　　　　可乐凤中翼

◎ 巩固练习 Exercise to Consolidate

根据前面所学，翻译菜名。(Translate the Chinese dishes into English according to what you have learnt and pictures above.)

冬菜扣肉＿＿＿＿＿＿＿＿＿＿＿　　腐乳猪蹄＿＿＿＿＿＿＿＿＿＿＿

爆炒牛肋骨＿＿＿＿＿＿＿＿＿＿＿　　番茄炖牛腩＿＿＿＿＿＿＿＿＿＿＿

芫爆散丹＿＿＿＿＿＿＿＿＿＿＿　　烤羊里脊＿＿＿＿＿＿＿＿＿＿＿

烤羊腿＿＿＿＿＿＿＿＿＿＿＿　　可乐凤中翼＿＿＿＿＿＿＿＿＿＿＿

鸿运蒸凤爪＿＿＿＿＿＿＿＿＿＿＿　　火燎鸭心＿＿＿＿＿＿＿＿＿＿＿

黑椒焖鸭胗＿＿＿＿＿＿＿＿＿＿＿　　葱爆肥牛＿＿＿＿＿＿＿＿＿＿＿

附图：

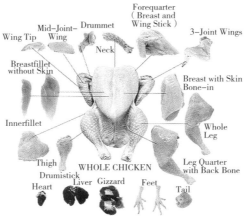

烹饪原料——水产品类 Cooking Material—Fishery Products

（1）海鲜（Seafood）

表 1

flatfish ['flæt,fiʃ] 比目鱼		flounder ['flaundə] 比目鱼、龙利 [c]	
cuttlefish ['kʌtl,fiʃ] 墨鱼		kelp [kelp] 海带	
ribbonfish ['ribənfiʃ] 带鱼		mackerel ['mækrəl] 马鲛鱼（单复数同形）	
turbot ['tə:bət] 多宝鱼（单复数同形）多宝鱼肉 [u]		squid [skwid] 鱿鱼	
prawn [prɔ:n] 大虾 [c]		shrimp [ʃrimp] 基围虾 [c]	
lobster ['lɔbstə] 龙虾 [c]；龙虾肉 [u]			

更多海鲜
扫描二维码获取

表 2

sea whelk	海螺 [c]	sea [si:] n. 海洋		whelk [welk] n. 螺	
sea moss	海苔 [u/c]			moss [mɔs] n. 苔藓	
sea bass	黑鲈			bass [beis] n. 鲈鱼	
small yellow croaker	小黄鱼 [c]	small [smɔ:l] adj. 小的	yellow ['jeləu] adj. 黄色的	croaker ['krəukə] n. 黄花鱼	
phoenix-tailed prawn	凤尾虾 [c]	phoenix ['fi:niks] n. 凤凰	tail [teil] n. 尾巴	prawn [prɔ:n] n. 明虾，对虾	

（2）河鲜（Fresh-water Products）

clam	蛤蚌 [c]；（供食用的）蛤肉 [u]	[klæm]	
cat fish	鲶鱼		
river crab	毛蟹，河蟹 [c]；蟹肉 [u]	river ['rivə] n. 河流	crab [kræb] n. 螃蟹 [c]；蟹肉 [u]

更多河鲜
扫描二维码获取

（3）其他（Others）（海鲜和河鲜均可表示）

eel	鳗鱼；鳝鱼 [c]	[i:l]	
sashimi	生鱼片	['sɑ:ʃimi]	
salmon	鲑鱼 [c]；鲑鱼肉 [u]	['sæmən]	
crab claw	蟹钳 [c]	claw [klɔ:] n. 爪，钳	
grass carp	草鱼（单复数同形）	grass [grɑ:s] n. 草	carp [kɑ:p] n. 鲤鱼
silver fish	银鱼	silver ['silvə] n. 银；银灰色	

更多其他例子
扫描二维码获取

美食欣赏 Enjoy the Delicacies

鲍汁豆腐　　　　　干煎带鱼　　　　　大蒜烧白鳝　　　　鲜豌豆炒河虾仁

◎ 巩固练习 Exercise to Consolidate

根据前面所学，翻译菜名。（Translate the Chinese dishes into English according to what you have learnt and pictures above.）

鲍鱼烧牛头_____　　鲍汁豆腐_____

干煎带鱼_____　　清蒸鳜鱼_____

豆腐烧鱼_____　　茄汁虾仁_____

生炒鳗片_____　　鲶鱼烧茄子_____

鲜豌豆炒河虾仁_____　　冬菜银鳕鱼_____

清蒸白鳝_____　　白灼生中虾_____

子姜虾_____　　大蒜烧白鳝_____

烹饪原料——蛋制品 Cooking Material—Egg Products

egg	蛋 [c]	[eg]
egg yolk	蛋黄 [u/c]	yolk [jəulk] n. 卵黄
egg-white	蛋白或蛋清	white [wait] n. 白色
omelet	煎蛋卷 [c]	['ɔmlit]

更多蛋制品
扫描二维码获取

◎ 巩固练习 Exercise to Consolidate

根据前面所学，翻译菜名。（Translate the Chinese dishes into English according to what you have learnt.）

蛋黄明虾_____　　木瓜牛奶蛋黄汁_____

皮蛋豆腐_____　　口蘑煎蛋卷_____

烹饪原料——主食和小吃 Cooking Material—Staple Food and Snacks

（1）稻米类（Rice）

porridge	粥		['pɔrɪdʒ]
black rice	黑米	black [blæk] *n.* 黑色	
rice flour	米粉	flour ['flauə] *n.* 面粉；粉状物质	
sticky rice	糯米	sticky ['stiki] *adj.* 黏性的	
Yuanxiao	元宵	也可称作 glutinous rice balls for Lantern Festival	Lantern Festival *n.* 元宵节
Tangyuan	汤圆	也可称作 glutinous rice balls	ball [bɔ:l] *n.* 球
Wotou	窝头		

更多稻米类
扫描二维码获取

（2）面食（Wheaten Food）

noodles	面条 [pl]		['nu:dl]
Youtiao	油条	也可称作 Deep-Fried Dough Sticks	dough [dəu] *n.* 生面团
cake	蛋糕；饼 [c]		[keik]
pancake	薄饼 [c]		['pænkeik]
bun	小圆面包 [c]		[bʌn]
spring roll	春卷 [c]	spring [spriŋ] *n.* 春天	roll [rəul] *n.* 卷
dumpling	饺子 [c]（Jiaozi）	['dʌmpliŋ] 还有汤团，面团布丁的意思	
Wonton	馄饨；云吞		['wɔn'tɔn]
Guotie	锅贴	= pan-fried meat dumplings	

更多面食
扫描二维码获取

（3）点心（Dim Sum）

dim sum	（汉）点心	dim [dim]	sum [sʌm]
puff	泡芙 [c]		[pʌf]
egg tart	蛋挞 [u/c]		tart [tɑːt] *n.* 果馅饼
muffin	小松饼 [c]		[ˈmʌfin]
toffee	太妃糖 [u/c]		[ˈtɔfi]

◎ 巩固练习 Exercise to Consolidate

① 写出下列主食和小吃的英文名。（Name these staple food and snacks in English.）

_____   _____   _____

② 根据前面所学，翻译菜名。（Translate the Chinese dishes into English according to what you have learnt.）

米饭_____   海皇炒饭_____

鸡汤面_____   红烧牛腩汤面_____

红烧排骨汤面_____   煎包_____

香滑芋蓉包_____   韭菜晶饼_____

萝卜丝酥饼_____   芋丝炸春卷_____

黑米小窝头_____   小米金瓜粥_____

绿豆粥_____   炸云吞_____

烹饪原料——水果类 Cooking Material—Fruits

apple	苹果 [c]	['æpl]
banana	香蕉 [c]	[bə'nɑːnə]
lemon	柠檬 [u/c]	['lemən]
litchi	荔枝	['litʃiː]
longan	龙眼	['lɔŋgən]

续表

pineapple	菠萝 [u/c]	[ˈpainˌæpl]
sugarcane	甘蔗	[ˈʃugəˌkein]
strawberry	草莓	[ˈstrɔːbəri]
watermelon	西瓜 [u/c]	[ˈwɔːtəmelən]
chestnut	栗子	[ˈtʃesnʌt]
snow pear	雪梨 [c]	pear [peə] n. 梨
kiwi fruit	猕猴桃	[ˈkiːwiː]
peanut	花生	[ˈpiːnʌt]
walnut	核桃	[ˈwɔːlnʌt]
date	枣子 [u/c]	[deit]

更多水果
扫描二维码获取

◎ 巩固练习 Exercise to Consolidate

① 用英文为下列水果命名。(Name the fruits below in English.)

② 将下列水果翻译成中文并试着画出这些水果。(Translate the fruits into Chinese and try to draw them by yourself.)

longan	pineapple	dragon fruit	kiwi fruit	water chestnut

③ 根据前面所学，翻译菜名。(Translate the Chinese dishes into English according to what you have learnt.)

明虾荔枝沙拉_____　　荔枝炒牛肉_____

淮山圆肉炖甲鱼_____　　菠萝虾球_____

冰梅凉瓜_____　　雪豆马蹄_____

松仁香菇＿＿＿＿＿＿＿＿＿＿＿＿＿＿　　板栗红烧肉＿＿＿＿＿＿＿＿＿＿＿＿＿＿

花生糕＿＿＿＿＿＿＿＿＿＿＿＿＿＿＿　　鸡丁核桃仁＿＿＿＿＿＿＿＿＿＿＿＿＿＿

雪梨炖百合＿＿＿＿＿＿＿＿＿＿＿＿＿　　杏仁炒南瓜＿＿＿＿＿＿＿＿＿＿＿＿＿＿

烹饪原料——干货制品 Cooking Material—Dried Products

更多干货制品
扫描二维码获取

英文	中文	音标/释义
sesame	芝麻	[ˈsezəmi]
bird's nest	燕窝	nest [nest] 巢，窝
deer antler	鹿茸	deer [diə] 鹿　　antler [ˈæntlə] 茸角
white fungus	银耳	white [wait] adj. 白色的　　fungus [ˈfʌŋgəs] 菌类 [u/c]（pl, fungi）
black fungus	黑木耳	black [blæk] adj. 黑色的
yellow fungus	黄耳	yellow [ˈjeləu] adj. 黄色的
dried radish	萝卜干	dried [draid] adj. 弄干了的　　radish [ˈrædiʃ] 萝卜 [c]
jellyfish	海蜇 [c]	[ˈdʒelifiʃ]
fish lip	鱼唇 [c]	fish [fiʃ] n. 鱼　　lip [lip] n. 嘴唇
fish maw	鱼肚	maw [mɔ:] n.（动物的）胃
sea cucumber	海参	cucumber [ˈkju:kʌmbə] n. 黄瓜
snow clam	雪蛤	snow [snəu] n. 雪　　clam [klæm] n. 蛤蚌 [c];（供食用的）蛤肉 [u]
lily bulb	百合 [c]	
black mushroom	香菇 [c]	
tea tree mushroom	茶树菇 [c]	详见烹饪原料——蔬菜类
wild mushroom	野菌菇 [c]	
straw mushroom	草菇 [c]	

美食欣赏 Enjoy the Delicacies

鲍汁扣花胶皇　　　枸杞蒸裙边　　　冰糖银耳炖雪梨

鲍汁扣辽参　　　鲜人参炖土鸡　　　燕窝鸽蛋

◎ 巩固练习 Exercise to Consolidate

根据前面所学，翻译菜名。(Translate the Chinese dishes into English according to what you have learnt and the pictures above.)

鲍汁扣辽参_____　　蛋花炒鱼肚_____

芝麻炸多春鱼_____　　木瓜腰豆煮海参_____

枸杞蒸裙边_____　　牛肝菌红烧豆腐_____

鲜人参炖土鸡_____　　燕窝鸽蛋_____

冰糖银耳炖雪梨_____　　萝卜干毛豆_____

白汁烧裙边_____　　鲍汁扣花胶皇_____

四、常用烹饪调料的英文表述
The English Expressions of Common Condiment

教学指引 Teaching Guideline

调料（即调味原料）在烹饪中虽然用量不大，却应用广泛，富于变化。在烹调过程中，调味原料的呈味成分连同菜点主配料的呈味成分一起，共同形成了菜点的不同风味特色。相关的词汇在菜名的表达中出现频率也很高，因此无论是从烹饪知识的角度，还是烹饪英语专业词汇的角度来说，学习烹饪调味原料的英文表述都有很大的意义，这也是我们把烹饪调料从烹饪原料中单独拿出来学习的原因。

知识背景介绍 The Introduction of Knowledge Background

调味原料的分类，具体如下。

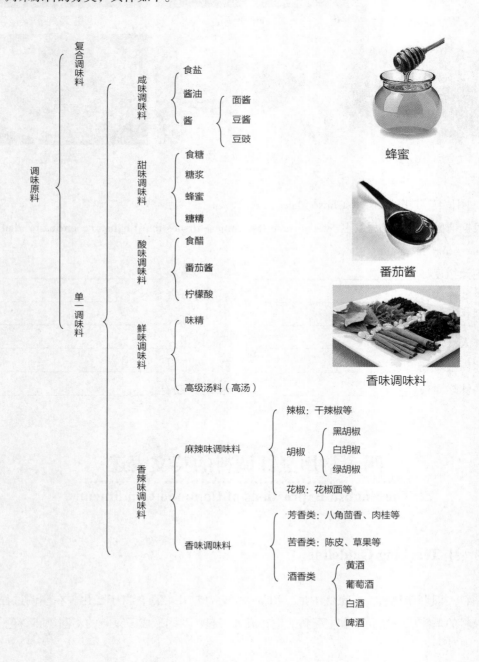

蜂蜜

番茄酱

香味调味料

学生学习 Student Learning

基础词汇学习 Basic Words Learning

salt [sɔːlt]	*n.* 盐
honey [ˈhʌni]	*n.* 蜂蜜
sugar [ˈʃugə]	*n.* 食糖
vinegar [ˈvinigə]	*n.* 食醋
pepper [ˈpepə]	*n.* 胡椒，辣椒
cumin [ˈkʌmin]	*n.* 小茴香，孜然
vanilla [vəˈnilə]	*n.* 香草
sauce [sɔːs]	*n.* 调味汁，少司
paste [peist]	*n.* 膏，糊状物
wine [wain]	*n.* 酒
spicy [ˈspaisi]	*adj.* 辛辣的，香的，多香料的

◎ 拓展学习 Extended Learning

rock candy =sugar candy =crystal sugar	冰糖	rock [rɔk] *n.* 岩石
		candy [ˈkændi] *n.* 糖果
		crystal [ˈkristəl] *n.* 水晶，结晶
aged vinegar	老醋	aged [ˈeidʒid] *adj.* 年老的

◎ 巩固练习 Exercise to Consolidate

为以下调料瓶贴上英文标签以方便外国客人使用。（Label these apothecary jars in English so that the foreign guests are able to use them freely.）

| 糖 | 盐 | 醋 | 胡椒 | 蜂蜜 |

常用调味汁和少司 The Common Sauce

soy sauce	酱油	soy [sɔi] 大豆

续表

brown sauce	棕色调味酱	brown [braun] adj. 棕色的
Maggi sauce	美极酱	Maggi [mædʒi] n. 美极（厨房调理食品品牌名）
ginger sauce	姜汁	ginger [ˈdʒindʒə] 姜
garlic sauce	蒜汁；蒜蓉	garlic [ˈɡɑːlik] 大蒜
curry sauce	咖喱汁	curry [ˈkəːri] 咖喱
rice wine sauce	香糟	rice wine 米酒
osmanthus sauce	桂花酱	osmanthus [ɔzˈmænθəs] 桂花
plum sauce	酸梅汁	plum [plʌm] 李子；梅子
saffron sauce	红花汁	saffron [ˈsæfrən] 藏红花
oyster sauce	蚝油	oyster [ˈɔistə] 蚝；牡蛎
Thai sauce	泰式酱	Thai [tai] adj. 泰国的
shrimp paste	虾酱	shrimp 虾
sesame paste	麻酱	sesame 芝麻
scallion oil	葱油	scallion 葱 oil [ɔil] 油

美食欣赏 Enjoy the Delicacies

冰糖甲鱼　　　　　冰花炖官燕　　　　　咖喱猪排饭

蒜蓉海带丝　　　　孜然寸骨　　　　　麻酱笋条

◎ 巩固练习 Exercise to Consolidate

根据前面所学，翻译菜名。（Translate the Chinese dishes into English according to what you have learnt.）

孜然寸骨＿＿＿＿＿＿＿＿＿＿＿＿　　香草蒜蓉炒鲜蘑＿＿＿＿＿＿＿＿＿＿

冰糖甲鱼＿＿＿＿＿＿＿＿＿＿＿＿　　冰花炖官燕＿＿＿＿＿＿＿＿＿＿＿

香酱爆鸭丝＿＿＿＿＿＿＿＿＿＿＿　　姜汁皮蛋＿＿＿＿＿＿＿＿＿＿＿＿

蒜蓉海带丝＿＿＿＿＿＿＿＿＿＿＿　　咖喱猪排饭＿＿＿＿＿＿＿＿＿＿＿

冰梅凉瓜＿＿＿＿＿＿＿＿＿＿＿＿　　蚝汁辽参扣鸭掌＿＿＿＿＿＿＿＿＿＿

葱油泼多宝鱼＿＿＿＿＿＿＿＿＿＿　　麻酱笋条＿＿＿＿＿＿＿＿＿＿＿＿

◎ 辣味调味品 Spicy Condiment

green pepper	绿胡椒；青椒	
bell pepper（sweet pepper）	甜辣椒	bell [bel] n. 钟状物 sweet [swi:t] adj. 甜的
hot pepper	辣椒；腌辣椒	hot [hɔt] adj. 热的；辣的
chili sauce	辣酱油	chili ['tʃili] n. 红辣椒；辣椒
spicy sauce	香辣汁；五香味	
hot spicy sauce	香辣汁；麻辣汁	
hot and sour sauce	酸辣汁	sour ['sauə] adj. 酸的
wasabi（mustard sauce）	芥末酱	wasabi [wa:'sa:bi] n. 芥末 mustard ['mʌstəd] n. 芥末

◎ 豆酱 Soy Bean Paste

soy bean paste	豆酱	soy bean 大豆
red bean paste	豆沙酱	red bean 红豆
traditional Beijing bean paste	老北京豆酱	traditional [trə'diʃənəl] adj. 传统的
sweet bean sauce	甜豆酱；京酱	
black bean sauce	豆豉汁	black bean 豆豉；黑大豆

酒香类 Wine

white wine	白葡萄酒
red wine	红葡萄酒
yellow wine	黄酒
Chinese Liquor	白酒（中式）
rice wine	米酒
aged rice wine	老酒
beer [biə]	啤酒

◎ 巩固练习 Exercise to Consolidate

① 在一场婚礼上，一个外国客人看到餐桌上的各种酒非常感兴趣。请向他介绍这些酒。(In the wedding ceremony, a foreign guest was curious about the various wines on the table. Introduce them to him.)

② 根据前面所学，翻译菜名。(Translate the Chinese dishes into English according to what you have learnt.)

青椒牛柳_____   辣白菜炒牛肉_____

酸辣瓜条_____   芥末木耳_____

京酱龙虾球_____   豆豉豆腐_____

红酒鹅肝_____   啤酒鸡_____

糟香鹅掌_____

五、常用菜肴风味的英文表述
The English Expressions of Common Flavors

教学指引 Teaching Guideline

"风味"在这里指的是一地特有的食品口味。通过不同的美味可以了解不同地方的特色菜肴。事实上，英文中对风味的表达比较简单，但是在学会这些表达之前，无论是教师还是学生都应该对其背景知识有一定的了解，学习烹饪英语不仅要学习烹饪语言知识，更要学习烹饪文化。

知识背景介绍 The Introduction of Knowledge Background

在我国，由于气候、地理、历史、物产及饮食风俗的不同，各地方的菜肴经过漫长的历史演变形成了一整套自成体系的烹饪技艺和风味，形成了各地区极具地方特色的菜系。一般来说，中国北方寒冷，菜肴以浓厚、咸味为主；华东地区气候温和，菜肴以甜味和咸味为主；西南地区多雨潮湿，菜肴多用麻辣浓味。目前，我国菜系的划分主要有"四大菜系""八大菜系""十大菜系"几种说法。现对"八大菜系"做一个简要说明。

"八大菜系"指的是：闽菜、鲁菜、川菜、粤菜、苏菜、浙菜、湘菜、徽菜。

（1）闽菜由福州、泉州、厦门等地方菜发展而成。以海味为主要原料，具有制作精细、色调美观、滋味清鲜的特点。口味以清淡甜酸为主，尤以"糟"味最具特色。烹调方法擅长炒、熘、煎、煨等。代表作品有"佛跳墙""鸡汤氽海蚌""龙心凤尾虾""淡糟香螺片""水晶干贝""红焖鲍鱼""红糟鸡丁"等。

（2）鲁菜又称山东菜，其特点是清香、鲜嫩、味纯，十分讲究清汤和奶汤的调制，清汤色清而鲜，奶汤色白而醇。代表作品有"符离集烧鸡""火腿炖甲鱼""腌鲜鳜鱼""火腿炖鞭笋""雪冬烧山鸡"等。

（3）川菜以"一菜一格，百菜百味"的鲜明个性著称于世。川菜以成都、重庆两地的菜肴为代表。其特点是十分注重调味，一般多用辣椒、花椒、胡椒、香醋、豆瓣、豆瓣酱等。口味有麻辣、酸辣、豆瓣、香豉、三椒、怪味等。烹调方法也颇具特色，擅长小煎、小炒、干烧、干煸。代表作品有"大煮干丝""怪味鸡块""麻婆豆腐"等。

（4）粤菜由广州、潮州、东江等地方菜发展而成。主要特点是选料精细、花色繁多、新颖奇异。口味以清淡、生脆爽口为主，还特别注重色、香、味、形俱佳，尤其讲究形态美观。代表作品有"烤乳猪""白灼虾""太爷鸡""黄埔炒蛋"等。

（5）苏菜主要由淮扬（淮安、扬州）、金陵（南京）、苏锡（苏州、无锡）、徐海（徐州、连云港）四大地方风味组成。其特点是选料严谨、制作精致、注意配色、讲究造型，菜肴有四季之别。烹调方法擅长炖、焖、烧、炒。菜肴口味清淡适口，甜咸适中，适应性强，南北皆宜。代表作品有"松子肉""枣方肉""百花酒焖肉""松鼠鳜鱼""糟煎白鱼""无锡脆鳝""清蒸鲥鱼"等。

（6）浙菜由杭州、宁波、绍兴等地的地方菜发展而成。具有清鲜、细嫩、制作精细的特点，擅长的烹调方法是爆、炒、烩、炸、烤、焖等。菜肴鲜美嫩滑、清爽不腻，色泽光润鲜艳。代表作品有"西湖醋鱼""龙井虾仁""叫化鸡""东坡肉""淡菜嵌肉""生爆鳝片""西湖莼菜汤"等。

（7）湘菜由湘江流域、洞庭湖区、湘西山区的地方菜发展而成。特点是用料广泛、油重色浓，制作上讲究原料的入味，口味注重辣酸、香鲜、软嫩适口。技法多样，尤重煨烤。代表作品有"麻辣子鸡""腊味合蒸""走油豆豉扣肉""银鱼火锅""红椒酿肉"等。

（8）徽菜由沿江、沿淮、徽州三个地区的地方菜发展而成。徽菜的特色是选料朴实，擅长烧、炖、蒸等烹调方法，菜肴具有"三重"的特点，即"重油""重酱色""重火功"。代表作品有"黄山炖鸡""火烘鱼""朱洪武豆腐""芙蓉蹄筋""御笔黄鳝"等。

学生学习 Student Learning

导入 Lead-in

与风味相关的菜肴表达在中文菜名中是有区别的，有的是一般菜名的叫法，比如"麻辣子鸡""生爆鳝片"等；有的是在菜名中直接体现出该菜所属的菜系，比如"上海春卷""湘味回锅肉"等，因此英文的菜名表达也会有所区别。

◎ 思考题 Questions

以下三个英文菜名的表述分别代表三种典型的有关菜肴风味的表达，仔细观察，结合导入部分的内容，看看你是否可以得出一些规律。

香辣猪扒：
Grilled Pork with Spicy Sauce

川味红汤鸡：
Chicken in Hot Spicy Sauce,
Sichuan Style

担担面：
Noodles, Sichuan Style

◎ 细解 Explanation

通过对上述中文菜名的分析，从风味角度来说，菜名的英文表达大致可分为以下两类。

① 普通表述，即不必特别体现出所属菜系的表述（Common Expressions）

例如：

香辣猪扒：Grilled Pork with Spicy Sauce

解析：Grilled（烹饪技法的过去分词）+ Pork（主料）+ with（连词）+ Spicy Sauce（汤汁）

蛋黄狮子头： Stewed Meat Ball with Egg Yolk

解析：Stewed（烹饪技法的过去分词）+ Meat Ball（主料）+ with（连词）+ Egg Yolk（辅料）

② 体现地方风味的表述（Expressions with Regional Flavors）

例如：

川味红汤鸡： Chicken in Hot Spicy Sauce，Sichuan Style

解析： Chicken（主料）+ in（连词）+ Hot Spicy Sauce（汤汁），Sichuan Style
　　　　　　　　　　　　　　　　　　　　　　　　　　　　　四川　风格

湘味回锅肉： Sautéed Pork with Pepper，Hunan Style

解析： Sautéed（烹饪技法的过去分词）+ Pork（主料）+ with（连词）+ Pepper（辅料），Hunan Style
　　　　　　　　　　　　　　　　　　　　　　　　　　　　　　　　　　　湖南　风格

北京炸酱面： Noodles with Soy Bean Paste，Beijing Style

解析： Noodles（主料）+ with（连词）+ Soy Bean Paste（辅料），Beijing Style
　　　　　　　　　　　　　　　　　　　　　　　　　　　　　北京　风格

提示 Tips

有些中文的菜名，虽然没有直接体现地方菜系，但是它本身的特点或者特色已经对其有所体现，我们把它称作隐性体现，这样的菜名在转换成英文表述时也要体现出地方菜系。

例如：

干烧牛肉：Dry-Braised Shredded Beef，Sichuan Style

解析： Dry-Braised（修饰词）+ Shredded（修饰词）+ Beef（主料），Sichuan Style
　　　　　　　　　　　　　　　　　　　　　　　　　　　　　四川　风格

担担面：Noodles，Sichuan Style

解析： Noodles（主料），Sichuan Style
　　　　　　　　　　　四川　风格

公式学习 Expressions Learning

◎ 思考题 Questions

我们已经学习了三个有关于菜肴表述的公式：

公式一（基础公式）：主料+（形状）+连词（and/with）+辅料+连词（in/with）+汤汁

公式二（拓展公式一）：烹饪技法+基础公式

公式三（拓展公式二）：修饰词+基础公式

根据上文讲述的风味命名规律，你是否可以总结出一个在我们学过的公式基础上的新公式呢？

◎ 拓展公式学习 Extended Expressions Learning

我们来看几个上文提到的例子：

北京炸酱面：Noodles with Soy Bean Paste, Beijing Style

解析：Noodles（主料）+ with（介词）+ Soy Bean Paste（辅料），Beijing Style
　　　 ─────────────基础公式───────────── + 北京 风格 Beijing Style

中式牛柳：Beef Filet with Tomato Sauce, Chinese Style

解析：Beef（主料）+ Filet（形状）+ with（连词）+ Tomato Sauce（辅料），Chinese Style
　　　 ─────────────基础公式───────────── + 中国 风格 Chinese Style

湘味回锅肉：Sautéed Pork with Pepper, Hunan Style

解析：Sautéed（烹饪技法的过去分词）+ Pork（主料）+ with（介词）+ Pepper（辅料），Hunan Style
　　　 ─────────────拓展公式一───────────── + 湖南 风格 Hunan Style

干烧牛肉：Dry-Braised Shredded Beef, Sichuan Style

解析：Dry-Braised（修饰词）+ Shredded（修饰词）+ Beef（主料），Sichuan Style
　　　 ─────────────拓展公式二───────────── + 四川 风格 Sichuan Style

由此我们得出了另外一个有关于菜名表述的公式——拓展公式三。

拓展公式三：（修饰词）+基础公式+ 地区/地方+Style
　　　　　　 ────拓展公式二────

提示 Tips

修饰词在括号内,表示其有可能被省略。

美食欣赏 Enjoy the Delicacies

潮州烧雁鹅　　　　　川汁牛柳　　　　　家常豆腐

台式蛋黄肉　　　　　上海油爆虾　　　　中式牛排

练习 Exercises

根据前面所学和以上图片,翻译菜名。(Translate the Chinese dishes into English according to what you have learnt and the pictures above.)

川汁牛柳 _____　　湖南牛肉 _____

潮州烧雁鹅 _____　　上海油爆虾 _____

上海菜煨面 _____　　日式蒸豆腐 _____

中式牛排 _____　　台式蛋黄肉 _____

家常豆腐 _____

六、菜名的英文表述小结
The Summary of the English Expressions of Chinese Dishes

教学指引 Teaching Guideline

　　本节的主要内容是在学习完烹饪原料、烹饪调料等各方面的词汇以后，更全面地学习关于菜肴表达的公式。当然这里学习的重点是根据菜名直接翻译的相关方法，有关意译的或者是直接用拼音翻译等的方法会在Extended Learning（学习拓展）环节做一个小结，让学生了解，但不要求掌握。上课过程中，教师一定要注重对词汇的学习和巩固，这是公式学习的基础。

学生学习 Student Learning

提示 Tips

　　我们已经学习了烹饪方法、烹饪原料、烹饪调料和风味的英文表达，词汇的学习是一个循序渐进的过程，千万不要急于求成，也不要被它吓倒，每天坚持一点，你会发现在掌握完这些词汇以后，再利用公式，大部分菜肴的表达已经难不倒你了，你已经掌握了烹饪英语中最重要也是最有难度的一个知识点了。

导入 Lead-in

◎ 思考题 Questions
回顾一下，到目前为止，我们学习的菜肴表达公式有哪些？

◎ 回顾 Review
① 基础公式：主料+（形状）+连词（and/with）+辅料+连词（in/with）+汤汁
② 拓展公式一：烹饪技法+基础公式
③ 拓展公式二：修饰词+基础公式
④ 拓展公式三：（修饰词）+基础公式 + 地区/地方+Style

公式学习 Expressions Learning

◎ 思考题 Questions

通过下述菜名以及所给提示，再结合学过的公式，试着推断一下将要学习的拓展公式。

砂锅萝卜羊排：Stewed Lamb Chops and Turnip in <u>Pottery Pot</u>
　　　　　　　　　　　　　　　　　　　　　　　　　砂锅

砂锅萝卜羊排

纸包鸡：Deep-Fried Chicken in <u>Tin Foil</u>
　　　　　　　　　　　　　　　　锡纸

纸包鸡

◎ 细解 Explanation

从以上例子中，我们可以看出，给出的菜名是在基础公式或者是拓展公式二的基础上多了器具的部分，由此得出另一个拓展公式。

◎ 拓展公式学习 Extended Expressions Learning

从上文的例子中我们可以得出拓展公式四：修饰词+基础公式+in+器具

砂锅萝卜羊排：Stewed Lamb Chops and Turnip in Pottery Pot

解析：Stewed（修饰词）+<u>Lamb Chops（主料）+ and +Turnip（主料）+in（介词）</u>+Pottery Pot（器具）
　　　　　　　　　　　　　　　　　　　　　　　　　　　　　　　　　　　　　砂锅

　　　　　　　　　　拓展公式一　　　　　　　　　　　+ Pottery Pot

表示器具的常用词

iron pot	铁锅	iron [ˈaiən] *adj.* 铁的	pot [pɔt] *n.* 壶；盆；罐
pottery pot	瓦罐；砂锅	pottery [ˈpɔtəri] *n.* 陶器	
stone pot	石锅	stone [stəun] *adj.* 石的；石制的	
tin foil	锡纸	tin [tin] *adj.* 锡制的	foil [fɔil] *n.* 箔；金属薄片

美食欣赏 Enjoy the Delicacies

砂锅白菜粉丝

砂锅鱼头豆腐

铁锅牛柳

石钵蟹黄豆腐

瓦罐山珍

练习 Exercises

根据前面所学和以上图片,翻译菜名。(Translate the Chinese dishes into English according to what you have learnt and the pictures above.)

砂锅白菜粉丝 _____　　石钵蟹黄豆腐 _____

砂锅鱼头豆腐 _____　　铁锅牛柳 _____

瓦罐山珍 _____

提示 Tips

① 菜名中的名词的单复数一般遵循:不可数名词用单数形式,可数名词用复数形式。

例如:五香云豆:Spicy Kidney Beans(Bean是可数名词,用复数)

白切鸡:Boiled Chicken with Sauce(Sauce是不可数名词,用单数)

② 整道菜中只有一件或者是用料过于细碎的,虽然是可数名词仍使用单数。

例如:砂锅鱼头豆腐:Stewed Fish Head with Tofu in Pottery Pot(Fish Head是

可数名词，但只有一件的，用单数）

韭菜炒河虾：Sautéed Shrimps（Shrimp是可数名词，用复数）with Leek（Leek是可数名词，太过细碎，用单数）

练习 Exercises

根据所给图片和提示，补充完整以下菜名，注意单复数。(Fill the blanks according to the pictures and clues given and pay attention to plural or singular form of the noun.)

川北凉粉
Clear_____in Chili Sauce

朝鲜辣白菜
Korean_____in Chili Sauce

陈皮兔肉
_____with Tangerine Flavor

夫妻肺片
_____in Chili Sauce

干拌牛舌
_____in Chili Sauce

红心鸭卷
Sliced_____with Egg Yolk

姜汁皮蛋
_____in Ginger Sauce

韭菜炒鸡蛋
_____with Leek

干拌顺风
_____in Chili Sauce

学习拓展
扫描二维码获取

项目二　厨房常用设备与工具
Kitchen Utensil and Cooking Equipment

一、刀具的英文表述
The English Expressions of Cutting Tools

教学指引 Teaching Guideline

中国菜肴品种繁多，菜式琳琅满目，其独到的烹饪精髓更是驰誉海内外。深受人们喜爱的菜肴，往往不仅营养丰富，而且形态可人。因此，烹饪文化既强调源远流长的烹饪技术，也追求精湛的刀工手法。正所谓"十分厨艺，七分刀工"，刀工技术在烹饪技术中起着不可忽视的作用。

说到刀工，自然就离不开刀具。本节的主要内容是对常用的烹饪刀具进行介绍，让学生了解并掌握相关的词汇，并在此基础上进行专业表述法的操练。在教学过程中，教师可以充分利用多媒体教学的优势，同时结合实物教学法，引导学生掌握相关词汇。

知识背景介绍 The Introduction of Knowledge Background

刀具，不仅仅体现出中西方饮食习惯的差异，也投射出不同的文化内涵。西方人认为刀尖向外而锋利更便于使用，中国人则觉得刀具毕竟是一种具有杀伤力的东西，内敛是中式刀的主题。菜刀、砍骨刀的刀面近于方形，而其他中式刀的刀背在末端呈弧形向下，刀尖几乎和刀刃保持在同一平面上。西式刀则相反，刀尖和刀背几乎在同一水平面，刀刃在末端呈弧形向上，刀尖直接向外。

整体上来说，西式刀比中式刀轻巧一些，功能上专刀专用，发展了一系列不同功用的小刀。两派的刀型不同，用法自然也有所不同。中式菜刀，是靠刀的重量，从上到下的切；西式刀比较轻，切法是刀尖几乎不离开案板，只是抬起刀的后半部分，像是铡刀的用法。

在烹饪中，刀具的分类可谓十分精细，本节主要介绍一些常用的刀具，要求学生掌握。中西餐刀具的分类如下。

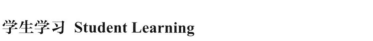

(1) 片刀　特点是重量较轻，刀身较窄而薄，钢质纯，刀刃锋利，使用灵活方便。适宜加工无骨无冻的动、植物性原料，但不适合加工硬性原料，主要用于加工片、条、丝、丁、米（粒）等形状，如片方干片、片肉片等。

片刀

(2) 切刀　形状与片刀相似，刀身比片刀略宽、略重、略厚，长短适中，应用范围广，既能用于切片、丝、条、丁、块，又能用于加工略带小骨或质地稍硬的原料。

(3) 砍刀　刀身较厚，刀头、刀背重量较重，呈拱形。根据各地方的特点，刀身有长有短，主要用于加工带骨、带冰或质地坚硬的原料，如猪头、排骨、猪脚爪等。

切刀

(4) 前切后砍刀　刀身大小与切刀相似，但刀的根部较切刀略厚，钢质如同砍刀，前半部分薄而锋利，近似切刀，特点是既能切又能砍。

学生学习 Student Learning

导入 Lead-in

砍刀

◎ 思考题 Questions

① 思考：常见的刀具有哪些？你能说出与刀具有关的单词吗？

② 讨论：根据你的生活经历和专业知识，你觉得刀具在烹饪中起着怎样的作用？不同的刀具分别有什么特征，应该如何使用？

不同种类的刀具 Words-Learning: Different Kinds of Knives

烹饪刀具品种繁多、用途各异，不同的刀具有各自的特点和功用。接下来，我们按照不同的标准对刀具进行简单归类，以便于大家理解和记忆。

（1）基础词汇 knife[刀]　复数形式：knives 厨房刀具的统称

（2）观察下列单词的特点，结合中文意思，总结出这一组刀具表达法的共性。

　　boning knife [去骨刀]　　　　　　carving knife [切肉刀]

　　paring knife [去皮刀]　　　　　　coring knife [去核刀]

　　slicing knife [片刀]　　　　　　　chopping knife [文武刀]

提示 Tips

① v-ing 动名词形式可以用来表示一种功能或用途。

② bone [骨头；去除骨头]；carve [切割]；pare [削，剥]；core [核心；挖去果核]；slice [切成薄片]。

◎ 巩固练习 Exercise to Consolidate

请根据图片写出相应的刀具。（According to the pictures below, try to write down the proper knives.）

（3）观察下列单词的特点，结合中文意思，总结出这一组刀具表达法的共性。

    meat knife [肉刀]　　　　　　　　fish knife [鱼刀]

    fruit knife [水果刀]　　　　　　　vegetable knife [蔬菜刀]

（4）观察下面这组单词，它们又有什么特点呢？它们与上一组有关系吗？

    bread knife [面包刀]　　　　　　　cheese knife [奶酪刀]

    tomato knife [番茄刀]　　　　　　 sandwich knife [三明治刀]

提示 Tips

观察可知，第三组的刀具均是对某一类食材刀具的总称，而第四类刀具则是针对某一类食材中具体某种食材的刀具。两组刀具的表达方式均为：名词 + knife。

试一试 Have a try

根据第三组刀具的表达规则，试着说出下列刀具的英文名。

[鸡蛋刀]　[黄油刀]　[马铃薯刀]

（5）个别刀具的表达

cleaver [劈刀] 一般是屠夫所用的切肉刀

ham slicer [火腿切片刀]

通过上述归类，我们可以发现，刀具的表达往往跟它们的作用和适用范围有关。所有的这些刀具，我们可以总称为 kitchen knife 或 chef knife，意为"厨房用刀"。

常用厨房刀具的英文表达 The Expressions of Kitchen Knives in Common Use

Chinese Style（中式）	Western Style（西式）
Slicing knife 片刀	Chef knife 西式厨刀
Carving knife 切刀	Bread knife 面包刀
Chopper 砍刀	Fillet knife 鱼片刀
Chopping knife 文武刀	Carving steel 磨刀棍

刀的不同部位 Different Parts of Knives

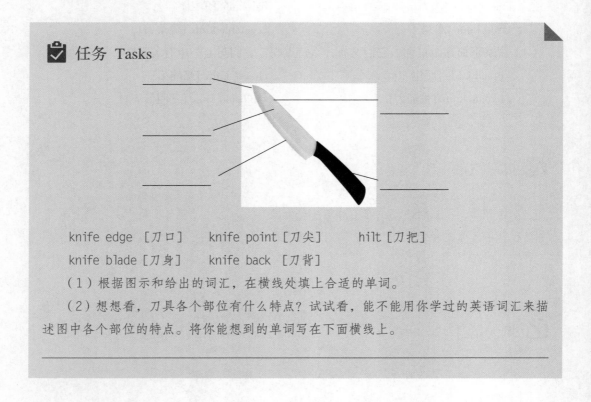

任务 Tasks

knife edge [刀口]　　knife point [刀尖]　　hilt [刀把]
knife blade [刀身]　　knife back [刀背]

（1）根据图示和给出的词汇，在横线处填上合适的单词。

（2）想想看，刀具各个部位有什么特点？试试看，能不能用你学过的英语词汇来描述图中各个部位的特点。将你能想到的单词写在下面横线上。

相关词汇 Related Words

了解并掌握有关刀具的相关知识和基本词汇，有助于我们更好地学习烹饪知识，掌握烹饪英语。当然，仅仅掌握刀具的名称词汇是不够的，我们需要了解与之相关的一些固定搭配，才能更好地运用相关知识，提高语言实践能力。接下来，让我们一起来了解一下与刀具有关的几个基本词汇。

（1）形容词

blunt [钝的]　　　　　　　　　　sharp [锋利的]

long [长的]　　　　　　　　　　short [短的]

two-edged [双刃的]　　　　　　single-edged [单刃的]

（2）动词

  cut [切]   chop [剁]   chip [削]

提示 Tips

  需要注意的是，这三个动词虽然都是将烹饪材料加工分解，但有一定的区别。cut 强调分解成块；chop 强调用力将食材加工成细小物，往往指馅类；chip 则是只加工成较小的片状或条状。联想一下，chip 作名词的时候就是"薯条"的意思。

需要掌握的字词 Words and Expressions You Need to Grasp

knife [naif] *n.* 刀   edge [edʒ] *n.* 边
cleaver [ˈkli:və] *n.* 切肉刀   ham [hæm] *n.* 火腿
vegetable [ˈvedʒitəbl] *n.* 蔬菜   point [pɔint] *n.* 尖，顶
kitchen [ˈkitʃin] *n.* 厨房   blade [bleid] *n.* 刀刃，刀片
chef [ʃef] *n.* 厨师   hilt [hilt] *n.* 刀把
sharp [ʃɑ:p] *adj.* 锋利的   blunt [blʌnt] *adj.* 钝的
chop [tʃɔp] *v.* 剁   chip [tʃip] *v.* 削
slice [slais] *v.* 切（片）

练习 Exercises

（1）根据图片写出恰当的动词。(Write down a proper verb according to the pictures.)

  _____   _____   _____

（2）根据所学知识，给图片中的食材选择合适的刀具，填在图片下方的横线上。(Choose a proper knife for each picture according to what have been learnt.)

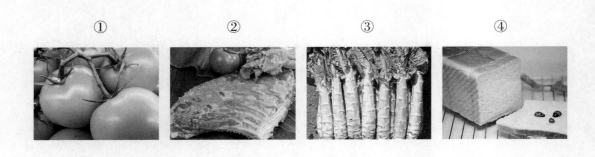

① _____　② _____　③ _____　④ _____

（3）练习（2）图中的四种食材，分别适合哪种加工方式？请为它们选择合适的动词。
（Which way can those food above be cut? Choose a proper verb for them.）

① _____　　　　　　　② _____
③ _____　　　　　　　④ _____

（4）猜猜下列图片是哪种刀。（Try to guess which kind of knife it is.）

_____

二、加热设备的英文表述
The English Expressions of Cookware

教学指引 Teaching Guideline

从本节开始，我们将介绍部分重要的厨房设备，帮助学生更全面地掌握烹饪英语知识。本节内容主要是加热设备的介绍。由于汉语字词和英语词汇并不是完全一一对应的，所以一些词汇存在着"一对多"或"多对一"的情况，在一定程度上给学生带来了一些困惑。在教学过程中，建议教师可以结合汉语和英语的特点，从两种语言的不同点出发，引导学生从语言性质和表达习惯的区别上去把握相关专业词汇的含义，以便能够举一反三、灵活掌握。

在教学过程中，教师可以充分利用多媒体教学的优势，同时结合实物教学法，提高学生兴趣，充分发挥学生的主观能动性，引导学生掌握教学内容。

知识背景介绍 The Introduction of Knowledge Background

在众多的厨房设备中，加热设备是不可或缺的重要部分。厨房加热设备，主要是指中、西餐及面点厨房各种热能的烹饪、蒸煮、烘烤等使菜点由生到熟、由原料到成品的制作设备。下面是常见的一些加热设备。

（1）中餐菜肴加热设备　主要包括煤气炉具（煤气炒炉、汤炉、煤气油炸炉等）、蒸汽炉具（蒸汽夹层锅、蒸柜等）。

（2）西餐菜肴加热设备　主要包括扒炉、电烤箱、电面火烤炉、西式煤气平头炉、电温藏箱、微波炉、电磁感应灶等。

（3）面点加热设备　主要包括饺子机、煤气蒸炉等。

各类加热设备都有它们各自的特点。了解每种设备的工作原理和特征，有助于更好地理解其英文表达。

学生学习 Student Learning

导入 Lead-in

◎ 思考题 Questions
① 记得那只可恶、可爱又可怜的灰太狼吗？灰太狼总是被什么打晕？
② 讨论：常见的烹饪设备有哪些？其中哪些属于加热设备？

语言知识 Language Knowledge

在中文当中，我们最常用的关于厨房加热设备的字词是"锅""炉""煲"等，比如炒锅、火锅、电炉、电饭煲等。观察一下我们会发现，汉字的表意多与偏旁有关，比如说被称作"锅"的多为金属，被称作"炉"和"煲"则常与火有关。随着电的发明使用和更多新技术的出现，一些汉字的意义与最初的含义有所变动，但是我们还是可以从文字的偏旁上去考虑它们的基本意义。

在英文当中，可不存在"偏旁"这样的情况，那么翻译的时候同样的"锅"，我们如何来选择不同的翻译法呢？首先，让我们一起来了解几个比较容易混淆的基础单词。

pot [罐，壶]   pan [平底锅，盘子]   boiler [汽锅]
stove [炉子] oven [烤箱，烤炉]

◎ 思考题 Questions

想想看，如何来区别上述几个单词的含义？

提示 Tips

从单词的最基本含义上来考虑。

◎ 巩固练习 Exercise to Consolidate

试着用刚学习过的单词命名下面的加热设备。（Try to name the cookware below according to what you have just learnt.）

各类烹饪加热设备 Different Kinds of Cookware

接下来，让我们一起来学习按照上面五个单词的含义来归类的五组加热设备。

（1）pot [罐，壶]　　　　　　　　　stew pot [蒸煮锅]
　　crock pot [瓦炖锅]　　　　　　　soup pot [汤罐]
　　coffee pot [咖啡壶]

◎ 细解 Explanation

pot：最基本的含义是"罐、壶"，理解的时候应该从设备的形状上来把握。有一定深度，且开口较小的，大部分可以称为pot。

（2）pan [平底锅，盘子]　　　　　　chip pan [油炸锅]
　　milk pan [炖奶锅]　　　　　　　baking pan [烘盘]
　　frying pan [油煎锅]

提示 Tips

回忆上一节内容中v-ing动名词形式表示用途的语言知识。

◎ 细解 Explanation

pan：有两个基本意思，一个是"平底锅"；另一个是"盘子"。与加热设备有关的大部分都是前一个意思，但也有一些是后一个意思。理解时注意两个要点：一是"平底"；二是"开口较大"。

（3）boiler [汽锅]　　　　　　　　　steam boiler [蒸锅]
　　double boiler [双层蒸锅]　　　　rice boiler [饭煲]
　　egg boiler [电煮蛋器]

◎ 细解 Explanation

boiler：最基本的含义是"汽锅"。我们可以从 boil 这个动词上去理解，boil是"用液体煮"的意思。因此，我们可以从加工时需不需"用液体煮"来把握。

试一试 Have a try

根据第三组设备的表达规则，试着说出下列设备的英文名。
[奶煲]　[鱼煲]　[汤煲]

（4）stove [炉子]　　　　spirit stove [酒精炉]　　　　electric stove [电炉]

◎ 细解 Explanation

stove：与以上三者相比，stove的特征是"相对封闭"，因为最初的炉子是一个封闭形、内中空的器具。

（5）oven [烤箱，烤炉]

大部分烘烤类的加热设施，在英文当中我们都可以统称为oven。

microwave oven [微波炉]

提示 Tips

掌握上述单词分析和例子学习，我们就能轻松区分pot, pan, boiler, stove和oven了。

◎ 巩固练习 Exercise to Consolidate

试着用刚学习过的单词再一次命名下面的加热设备。（Try to name the cookware below again according to what you have just learnt.）

_____　_____　_____　_____

_____　_____　_____　_____

_____　_____　_____　_____

◎ 拓展学习 Extended Learning

除了上面提到的几组加热设备，以下几种常见的设备也需要了解。

（6）some other cookware

pressure cooker [高压锅]　　　　gas range [煤气灶]　　　　electric range [电灶]

automatic rice cooker [电饭煲]　　chafing dish [火锅]

讨论 Discussion

如何记忆上述几种设备？

常用厨房加热设备的英文表达 The Expressions of Kitchen Cookware in Common Use

Chinese Style（中式）	Western Style（西式）
gas frying stove 煤气炒炉	griddle 扒炉
soup stove 汤炉	electric oven 电烤箱
fryer 油炸炉	termotank 保温箱
steam jacket cooker 蒸汽夹层锅	microwave oven 微波炉
steam cabinet 蒸柜	induction cooker 电磁炉

蒸柜

微波炉

提示 Tips

有很多加热设备都分为二头、三头、四头等。在英文中用"数字-headed"表示。例如，"双头炒炉"可以表示为Two-headed frying stove，"四头汤炉"可表示为Four-headed soup stove。

需要掌握的字词 Words and Expressions You Need to Grasp

pot [pɔt] n. 罐，壶　　　　　　　oven ['ʌvn] n. 烤箱，烤炉

pan [pæn] n. 平底锅，盘子　　　　microwave ['maikrəweiv] n. 微波

boiler [ˈbɔilə] *n.* 锅　　　　　　spirit [ˈspirit] *n.* 酒精

stove [stəuv] *n.* 炉子　　　　　chafe [tʃeif] *v.* 摩擦

electric [iˈlektrik] *adj.* 电的　　automatic [ˌɔːtəˈmætik] *adj.* 自动的

bake [beik] *v.* 烘　　　　　　　fry [frai] *v.* 油炸

练习 Exercises

（1）根据所学知识，给图片中的食材选择合适的加热设备，填在图片下方横线上。（Find the proper cookware for each picture according to what have learnt.）（答案不唯一）

① ② ③ ④

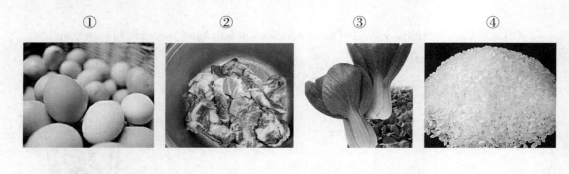

_____　_____　_____　_____

（2）回忆一下，最常用的加热设备有哪些？怎样归类？常用的英文表达是什么？（Try to think about the different kinds of cookware in daily use? How to classify them? How to say them in English？）

三、盛器的英文表述
The English Expressions of Containing Equipment

教学指引 Teaching Guideline

中国烹饪历来讲究美食美器。一道精美的菜点，如能盛放在与之相得益彰的盛器中，则更能展现出菜点的色、香、味、形、意来。本节我们将对厨房盛器进行基本介绍，让学生对盛器的英文表达有个初步了解，并在此基础上学习和运用相关英语语言知识。通过前两节内容的学习，学生已经掌握了一部分厨房用具的基本语言知识，对相关词汇的分类和使用技巧有了初步认识。在本节课的教学过程中，教师可以通过课堂任务的灵活设置，引导学生对前两节内容进行复习，并在此基础上导入新课。在学习过程中，应指导学生举一反三，结合已学知识来学习新知识，从而达到事半功倍的英语学习效果。其次，在本节的教学过程中，可以让学生对语言知识进行综合操练，结合新词汇的学习，巩固原有的词汇知识。

知识背景介绍 The Introduction of Knowledge Background

众所周知，菜肴制成后，都要用盘、碗盛装才能上席食用。值得注意的是不同的盛器对菜肴有着不同的作用和影响。合适的盛器，可以把菜肴衬托得更加美观，给人以悦目的感觉。

菜肴装盘时所用的盛器样式很多，规格大小不一，且在使用上各地也有所不同，常见的有以下几种。

（1）腰盘（长盘） 尺寸大小不一，最小的长轴5.5寸，最大的长轴21寸，小的可盛饭菜，大的多作盛鸡、鸭、鱼及筵席冷盘之用。

（2）圆盘 圆形，最小的直径5寸，最大的直径16寸，用途与腰盘相同。

（3）汤盘 盘底较深，最小的直径6寸，最大的直径约12寸，主要用于盛装烩菜或汤汁较多的菜。有些分量较多的炒菜如鳝糊往往也用汤盘。

（4）汤碗 汤碗是专作盛汤之用，直径一般为5~12寸。另外还有一种有盖的汤碗，叫碗品锅，盛整只鸡、鸭等汤菜用。

（5）扣碗 扣碗专用于盛扣肉、扣鸡、扣鸭等，直径一般为5~8寸。另外还有一种扣钵，一般用来盛全鸡、全鸭、全蹄等。

（6）砂锅 砂锅既是用来加热的用具，又是上席的盛具。特点是散热慢，故适用于煨、焖等需要用小火加热的烹调方法制成的菜肴。

学生学习 Student Learning

导入 Lead-in

◎ 思考题 Questions
① 常见的厨房盛器有哪些？它们分别有什么样的特征？
② 你能说出上图中的盛器吗？讨论一下哪些学过的英语词汇可以用来描述图中的盛器？

基础知识 Basic Knowledge

厨房盛器种类丰富，除了常用的"盘、碗、碟、盆"等用于盛菜的用餐盛器，还有各类饮品盛器和调味品盛器等。本节课，我们重点向大家介绍用餐盛器和饮品盛器。

各类用餐盛器 Different Kinds of Dishware

最常见的三类用餐盛器：plate [浅盘，碟]   dish [深盘]   bowl [碗]

plate   dish   bowl

◎ 思考题1 Question1

结合你的生活经验想想看，这三种盛器有什么主要区别？

① plate [浅盘，碟]

按照形状大小分

round plate [圆盘]   square plate [方盘]   oval plate [椭圆盘]

platter [大浅盘]   medium platter [中长盘]   small platter [小长盘]

总结：当按照形状大小来命名盛具的名称时，通常是"形容词+名词"形式的一个词组。

> 📚 **提示 Tips**
>
> plate 和 platter 的共同点和区别如下。
>
> 二者的共同点：都是浅盘。
>
> 二者的区别：主要体现在大小上，相对于 plate 而言，platter 往往是指较大的浅盘。

按照用途分

fruit plate [果盘]   dinner plate [（西餐用）大盘]   dessert plate [点心碟]

place plate [底盘]　　　　　food platter [大餐盘]　　　　salad plate [沙拉盘]
bread plate [面包盘]

总结：当按照用途来命名盛具的名称时，通常是"名词1+名词2"形式的一个词组。

② dish [深盘]
soup dish [汤盆]　　　　　fruit dish [果碟]　　　　　　cake dish [糕饼碟]
covered candy dish [有盖糖果碟]

③ bowl [碗]
soup bowl [汤碗]　　　　　cereal bowl [谷类食用碗]　　　dessert bowl [甜点碗]

④ 除此之外，还有一些不同表达法的盛具
例如：tray [托盘]　jar [盅]　pot [罐]
相应的"名词1+名词2"词组有relish tray [调味品托盘]　biscuit jar [饼干盅]　soup pot [汤罐] 等。

◎ 思考题2 Question2

想想看，这里的soup pot [汤罐] 和上面的soup bowl [汤碗]两者在意思上有什么联系，又有什么不同？你能结合自己的实践经验举出另外的例子吗？

总结："名词1+名词2"形式的词组中，同样的食材放在不同的盛具中，名词1不变名词2变；不同的食材放在相同的盛具中，名词2不变名词1变。

各类饮品盛器 Different Kinds of Drink-ware

（1）cup类型的
coffee cup [咖啡杯]　　　sake cup [（日本）清酒杯]　　tea cup [茶杯]
cream soup cup [奶汤杯]

（2）glass类型的
juice glass [果汁杯]　　　iced tea glass [冰茶杯]　　　wine glass [葡萄酒杯]
Whiskey glass [威士忌小酒杯（盛放烈酒）]

◎ 思考题 Questions

想想看：我们在前面提到的"名词1+名词2"词组的规律，在这里是不是也适用呢？

（3）其他基本类型
goblet [高脚杯]　　mug [马克杯]　　tankard [大啤酒杯]　　covered can [有盖大杯]

◎ 巩固练习 Exercise to Consolidate

根据图片中客人所点的饮料选择正确的饮品盛器并写出其英文名。（Choose the right

drink-ware for the guests according to what they've ordered in the following pictures and then write down the English name for each drink-ware.）

① ② ③

a) _____ b) _____ c) _____

常用厨房盛器的英文表达 The Expressions of Kitchen of Containing Equipment in Common Use

Chinese Style（中式）	Western Style（西式）
round plate [圆盘]	dinner plate [（西餐用）大盘]
place plate [底盘]	salad plate [沙拉盘]
soup dish [汤盆]	bread plate [面包盘]
cereal bowl [谷类食用碗]	dessert bowl [甜点碗]
tea cup [茶杯]	Whiskey glass [威士忌小酒杯（盛放烈酒）]
juice glass [果汁杯]	wine glass [葡萄酒杯]

需要掌握的字词 Words and Expressions You Need to Grasp

platter [ˈplætə] n. 大浅盘
dessert [diˈzə:t] n. 甜点
cereal [ˈsiəriəl] n. 谷物食品
relish [ˈreliʃ] n. 调味品
square [skwɛə] adj. 方的
oval [ˈəuvəl] adj. 椭圆的

salad [ˈsæləd] n. 沙拉
tray [trei] n. 托盘
jar [dʒɑ:] n. 盅
biscuit [ˈbiskit] n. 饼干
medium [ˈmi:diəm] adj. 中等的
sake [seik] n. 日本清酒

Whiskey ['wiski] *n.* 威士忌　　　　juice [dʒu (:) s] *n.* 果汁

练习 Exercises

（1）根据本章所学知识，写出图片中食材的名称，并结合专业知识为它们选择合适的加工刀具、加热设备和盛器。将答案填写在图片下方表格中。(Name the food below and find the proper knife, cookware and dishware for each food according to what you have learnt.)（答案不唯一）

① ② ③ ④

图片	食材	加工刀具	加热设备	盛器
①				
②				
③				
④				

（2）根据下面给出的名词，从Group1和Group2中各选一个单词，按照"名词1+名词2"词组的规律，拼凑词组。(Choose one word from Group1 and another word from Group2, and then make the new word according to the rule.)（答案不唯一）

Group1：plate　platter　dish　bowl　tray　jar　pot
Group2：dinner　cake　soup　dessert　salad　square

_____
_____
_____
_____

四、辅助设备的英文表述
The English Expressions of Auxiliary Equipment

教学指引 Teaching Guideline

　　有效、齐全的辅助设备在厨房设施中是不可或缺的。在前三节的教学内容中，我们已经介绍了刀具、加热设备和盛器的相关知识，并在此基础上进行了一定的语言练习。在本节的教学过程中，我们将对厨房辅助设备进行简要介绍，以完善学生的知识结构并进一步提高语言运用能力。

　　本节学习任务相对较轻，因此在本节内容的教学过程中，教师可以结合课堂实际情况，将前面三节课程的内容融合进去，进行多层次的操练和运用。另外，教师也可以根据学生的掌握情况进行复习，并在此基础上设计可行性任务，引导学生运用语言，达到活学活用的目的。教师也可根据学生的兴趣爱好及学习能力，适当地补充课外知识，提高学生的语言综合能力，并加强他们的自主学习能力。

知识背景介绍 The Introduction of Knowledge Background

　　厨房的辅助设备主要是指加工、贮藏、烹调以外，与厨房生产关系也很密切的一些工具。主要有洗涤设备、备餐设备和抽排油烟设备等。

　　洗涤设备主要是指配合和满足厨房生产和厨房服务需要，餐饮企业配备和设置的洗碗、消毒、餐具保养、贮藏等相关工具。主要包括洗碟机、容器清洗机、银器抛光机、高压喷射机和餐具消毒柜。

　　（1）洗碟机又称洗碗机　有的是单体小型洗碟机，有的是与水槽废肴处理机结合在一起的组合式洗碟机，还有大型传输型洗碟机。

　　（2）容器清洗机　是专门清洗较大容器的洗涤机。

　　（3）银器抛光机　银餐具使用一段时间后表面会产生一层黑色的氧化层，使银器失去光泽，影响美观。银器抛光机利用容器内的小钢珠与银餐具一起翻滚，借助银珠与银器的摩擦除去银餐具表面的斑迹。

　　（4）高压喷射机　是一种多用途的洗涤设备，能喷出高压的热水，水温可以调节并能自动加入清洁剂。

　　（5）餐具消毒柜　大小不一，常见的有直接通气式和远红外加热式两种。

　　（6）备餐设备　是指配备在备餐间，以方便服务员进行备餐服务的设备，主要有电热开水器、全自动制冰机等。

（7）抽排油烟设备　　主要指用于将厨房烹调时产生的烟汽及时抽排出厨房的各类烟罩等，这些设备及其正常的运行是保证厨房良好空气的基础。抽排油烟设备，最简单的有排风扇，其特点是设备简单、投资少、排风效果较好，但容易污染环境；另外有滤网式烟罩，排气效果好，排油烟也可，但清洗工作量大。比较先进的抽排油烟设备是运水烟罩。

学生学习 Student Learning

导入 Lead-in

◎ 思考题 Questions
① 想想看，常见的厨房辅助设备有哪些？它们有什么作用？
② 讨论：结合已有知识和实践经验，你觉得目前的厨房辅助设备是否齐全？

基础知识 Basic Knowledge
厨房辅助设备主要可分为以下几类：洗涤设备、备餐设备、抽排油烟设备和食材加工设备。接下来，我们学习厨房辅助设备的相关知识。
（1）洗涤设备（Washing Equipment）

◎ Group 1
dishwasher（洗碗碟机）　　　　　　　　automatic dryer（自动干燥机）
container cleaning machine（容器清洗机）　　burnishing machine（银器抛光器）
high pressure spray washing machine（高压喷射机）
tableware disinfecting cabinet（餐具消毒柜）

◎ Group 2
kitchen sink（洗涤盆）　　mop（洗碗刷）　　bottle brush（洗瓶刷）
dish cloth（抹碗布）　　　dust cloth（抹布）　　detergent（洗涤剂）

◎ 细解 Explanation

Group 1 是常见的大型洗涤设备，而 Group 2 则是小型清洁器具。

◎ 巩固练习 Exercise to Consolidate

Match Task

下列洗涤设备，哪些比较适合洗瓶子，哪些比较适合洗碗？请找到符合条件的气球，将其线连到瓶子或碗上，有些气球可以同时被连。

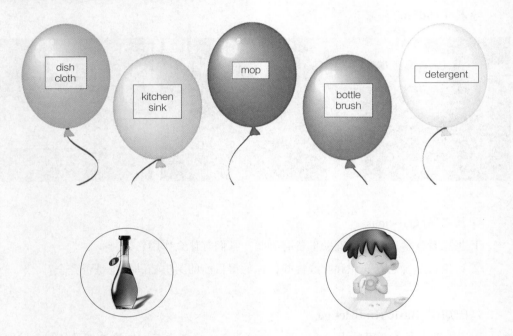

（2）备餐设备（Food Preparing Equipment）

cabinet（橱柜）

refrigerated cabinet（冷藏柜）

freezer（冷藏器）

freezing compartment（冷冻间）

defroster（除霜器）

ice cube keeper（储冰块器）

ice breaker（碎冰器）

ice machine（制冰机）

automatic electric water boiler（全自动电热开水器）

boiling plate（电热器）

temperature control switch（温度调节器）

◎ 细解 Explanation
观察可知，备餐设备一般有两个用途：冷藏和保温。

◎ 巩固练习 Exercise to Consolidate
Match Task
过年时，你们家买了一箱的虾，但是实在太多了，一家人吃不掉。可以放进下列哪些设备中？请将线连上。

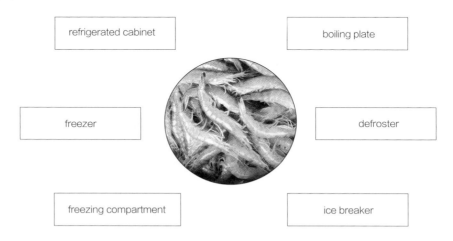

（3）抽排油烟设备（Oil Smoker Exhausting Equipment）
motor-operated fan（电动排风扇）　　　　cooker hood（排油烟机）
petticoat pipe（烟罩）　　　　　　　　　　hydro-vent system（运水烟罩系统）

（4）食材加工设备（Ingredients Preparing Equipment）
meat grinder（绞肉机）　　electric juicer（电榨汁机）　　mixer（搅碎机）
blender（搅拌器）　　　　peeling mill（去壳器）　　　　noodle press（制面机）
food slicer（切片机）　　　food chopper（切碎机）　　　electric grinder（电磨机）
potato masher（马铃薯捣烂器）

◎ 细解 Explanation
食材加工设备主要有"机"和"器"两大类，但是在学习英语词汇的时候要注意，不同的"机"和"器"所用的词汇大不一样。

◎ 巩固练习 Exercise to Consolidate
Match Task
在箭头上写出所需设备的英文名，以达到从左图到右图的效果。

常用厨房辅助设备的英文表达 The Expressions of Auxiliary Equipment in Common Use

dishwasher 洗碗碟机
tableware disinfecting cabinet 餐具消毒柜
automatic dryer 自动干燥机
cabinet 橱柜
freezing compartment 冷冻间
cooker hood 排油烟机

需要掌握的字词 Words and Expressions You Need to Grasp

dishwasher [ˈdɪʃwɒʃə(r)] n. 洗碗碟机
dryer [ˈdraɪər] n. 干燥机
tableware [ˈteɪblweə] n. 餐具
detergent [dɪˈtəːdʒənt] n. 洗涤剂
automatic [ˌɔːtəˈmætɪk] adj. 自动的
ingredient [ɪnˈɡriːdiənt] n. 原料
disinfect [ˌdɪsɪnˈfekt] vt. 消毒，杀菌

cabinet [ˈkæbɪnɪt] n. 橱柜
grinder [ˈɡraɪndə] n. 研磨器
machine [məˈʃiːn] n. 机器
mixer [ˈmɪksər] n. 搅碎机
hood [hʊd] n. 排风罩
ice breaker n. 碎冰器

练习 Exercises

（1）根据所学知识，在横线上写出图片中设备的英文名。（Name the equipment below according to what you have learnt.）

（2）结合本节内容，重新谈谈你对厨房辅助用具的看法。它们发挥着什么样的作用？
（Try to talk about your new ideas about Auxiliary Equipment and the function according to what you have learnt in this section.）

项目三　厨师岗位英语
English for the Cook on the Post

教学指引　Teaching Guideline

　　酒店内餐饮部厨房是否运作良好关乎整个酒店餐饮质量的优劣，所以清楚地明白厨房岗位设置以及相关职责十分重要，而厨房正常的运作除了各司其职、明确岗位责任之外，还要保证厨房运作的安全卫生，这是该专业每个学生都要明白的，也是在专业英语中必须了解和掌握的。在烹饪专业理论课上，专业老师会详细介绍厨房岗位设置和分工以及厨房安全要注意的方方面面，学生在实习的时候也会轮岗体验不同的厨房岗位。老师在上课的时候可以结合学生已经学过的专业知识以及学生的实习体验，让学生了解和掌握厨师岗位的英语表达，并为职业模块中相关对话学习打好词汇基础，同时在上课过程中通过对厨师岗位英语的学习，也要让学生进一步认识到不同岗位有不同的分工，要成为一个完整团队，合作才是最重要的。

知识背景介绍 The Introduction of Knowledge Background

要学生弄清楚酒店厨房岗位设置以及岗位设置在整个酒店运作时所处的位置，这样可以让学生重温专业理论知识，同时可以顺利导入本课所学内容。不同酒店的部分设置会略有出入，大致分工见下图。

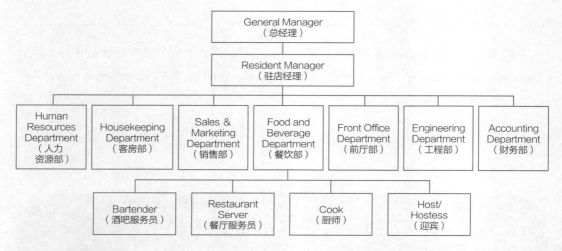

为了保障厨房的卫生与安全，常见的厨房管理方法主要有五常法和4D现场管理体系。

五常法也称5S现场管理法，指的是整理、整顿、清扫、清洁、素养。现在更多厨房都引进了7S或者8S管理法，比如8S管理法主要指整理、整顿、清扫、清洁、素养、安全、节约、学习。

4D现场管理体系又称卓越现场管理法，4D具体是：1D是整理到位，2D是责任到位，3D是执行到位，4D是培训到位。

一、厨房岗位的英文表述
The English Expressions of the Post in the Kitchen

学生学习 Student Learning

导入 Lead-in

◎ 思考题 Questions

① 回顾酒店的部门设置，再想一想厨房内部各个岗位

是如何设置的。
② 每个酒店的厨房岗位设置的差异主要原因是什么？

◎ 基础知识 Basic Knowledge

厨房岗位分工繁多，各司其职又相辅相成。接下来，我们来学习厨房各个岗位的英文表达。

（1）主厨 Chef

Executive Chef	行政总厨	Assistant Executive Chef	行政副总厨
Sous-chef	助理厨师长	Larder Chef	红案厨师长
Pastry Chef	白案厨师长		

（2）厨师 Cook

Sauce Cook	调料厨师	Soup Cook	汤料厨师
Roast Cook	烤肉厨师	Grill Cook	烧烤厨师
Breakfast Cook	早餐厨师	Night Cook	晚班厨师
Staff Cook	员工餐厨师	Relief Cook	替班厨师
Fish Cook	鱼菜厨师		

（3）其他岗位 Others

Pantryman	司膳总管	Caller	喊菜员
Commis	学徒工	Steward	膳务员
Dishwasher	洗碗工	Kitchen Porter	厨房搬运工

需要掌握的字词 Words and Expressions You Need to Grasp

executive [ɪgˈzekjətɪv] *adj.* 有执行权的　　chef [ʃef] *n.* 主厨；厨师
assistant [əˈsɪstənt] *n.* 助理　　　　　　sous [suː] *n.* 担任助理的
larder [ˈlɑːdə(r)] *n.* 肉贮藏处　　　　　relief [rɪˈliːf] *n.* 换班者
commis [kəˈmiːs] *n.* 厨助；学徒工　　　steward [ˈstjuːəd] *n.* 膳务员
dishwasher [ˈdɪʃwɒʃə(r)] *n.* 洗碗工　　　porter [ˈpɔːtə(r)] *n.* 搬运工

练习 Exercises

根据所学内容，看看以下原材料或者物品由厨房哪个岗位的厨师负责。（According to what you have learnt, find out who is in charge of these materials or food below.）

二、厨师职责的英文表述
The English Expressions of the Responsibilities of a Cook

学生学习 Student Learning

导入 Lead-in

◎ 思考题 Questions

想想看，作为烹饪专业的学生，你觉得厨师的岗位职责是什么？维护的是什么？

基础知识 Basic Knowledge

每一个厨师的职责归纳起来就是保证自身、食品材料和制作的卫生，厨房的卫生和安全，最终保证菜品的卫生和安全，确保酒店餐饮服务质量。接下来，我们来学习厨师基础职责的英文表述。

厨师基础职责 Basic Rules of the Duties as a Cook

（1）个人卫生 Personal Hygiene

职责1：保持头发干净并戴厨师帽。Duty 1: Keep hair clean and covered.

职责2：保持工装干净整洁。Duty 2: Keep uniforms clean and pressed.

职责3：保持指甲时常修剪。Duty 3: Keep fingernails short and well maintained.

职责4：常洗手，保持双手清洁。Duty 4: Wash hands frequently to keep hands clean.

职责5：用绷带包扎好伤口。Duty 5: Cover cuts with bandages.

（2）食品卫生 Food Hygiene

职责1：用干净的流水冲洗。Duty 1: Use clean, running water to rinse.

职责2：所有水果和蔬菜都要清洗。Duty 2: Wash all fruits and vegetables.

职责3：生熟食物分开放置。Duty 3: Keep raw food separated from cooked food.

职责4：熟食放在冰箱上层，生食放在冰箱下层，或者分冰箱放。Duty 4: Cooked above, raw below or separated refrigerators.

职责5：使用勺子或者钳子取食物。Duty 5: Use scoops or tongs for food.

（3）厨房安全 Kitchen Safety

职责1：厨房严禁吸烟。Duty 1: No smoking in the kitchen.

职责2：炉灶与可燃物之间应保持安全距离。Duty 2: Keep the oven from the inflammable goods in a safe distance.

职责3：炉具使用完毕，立即熄灭火焰，关闭气源，通风散热。Duty 3: Turn off the flame and gas as soon as dish is cooked. Keep the kitchen ventilated.

职责4：及时清理厨房设备，避免因油垢堆积而引起火灾。Duty 4: Clean the kitchen equipment timely in case of fire caused by greasy dirt.

职责5：离开厨房前确保厨房断电、断气、断火。Duty 5: Make sure all the power supplies, gas and flames are cut off before leaving the kitchen.

职责6：厨房应配备消防装置，厨房每人要熟悉报警程序和消防设施的使用。Duty 6: Fire equipment should be equipped in the kitchen. Everyone in the kitchen should be familiar with the procedure of giving a fire alarm and usage of fire equipment.

需要掌握的字词 Words and Expressions You Need to Grasp

hygiene [ˈhaɪdʒi:n] *n.* 卫生

press [pres] *v.* 压平

maintain [meɪnˈteɪn] *v.* 维持

rinse [rɪns] *v.* 冲洗

scoop [sku:p] *n.* 铲子；勺

inflammable [ɪnˈflæməbl] *adj.* 易燃的

greasy [ˈgri:zi] *adj.* 油腻的

procedure [prəˈsi:dʒə(r)] *n.* 程序

uniform [ˈju:nɪfɔ:m] *n.* 制服

fingernail [ˈfɪŋgəneɪl] *n.* 手指甲

bandage [ˈbændɪdʒ] *n.* 绷带

raw [rɔ:] *adj.* 生的

tong [tɒŋ] *n.* 钳子

flame [fleɪm] *n.* 火焰

power [ˈpaʊə(r)] *n.* 电力

usage [ˈju:sɪdʒ] *n.* 用法

练习 Exercises

（1）给下面几个职责归类。(Classify these rules of duties.)

- Personal Hygiene（个人卫生）
- Food Hygiene（食品卫生）
- Kitchen Safety（厨房安全）

① Keep the oven from the inflammable goods in a safe distance.

② Keep fingernails short and well maintained.

③ Use scoops or tongs for food.

（2）根据所学内容填空。(Fill in the blanks according to what you have learnt.)

| maintained | raw | as soon as | hair | rinse |

① Keep _____ clean and covered.

② Keep fingernails short and well _____.

③ Turn off the flame and gas _____ dish is cooked. Keep the kitchen ventilated.

④ Use clean, running water to _____.

⑤ Cooked above, _____ below or separated refrigerators.

三、厨房安全与卫生的英文表述

The English Expressions of Kitchen Safety and Hygiene

学生学习 Student Learning

导入 Lead-in

◎ 思考题 Questions

① 如何保证厨房的安全与卫生？

② 列举常见的酒店厨房管理方法。

基础知识 Basic Knowledge

酒店厨房常见的基础管理方法有五常法和4D现场管理体系。接下来，我们结合两种基础管理方法来学习厨房安全与卫生的管理方式。

（1）整理出厨房不必要的物品，将厨房物品分组并贴上标签。

Rule 1: Clean up the unnecessary things, divide goods into groups and label them.

（2）所有岗位权责明确。

Rule 2: Make the duties of all the positions clear.

（3）通过各种培训形式，确保员工将厨房条例谨记于心。

Rule 3: Make sure all staffs learn the rules in the kitchen by heart in various forms of trainings.

（4）所有条例必须执行到位。

Rule 4: All the rules must be carried out strictly.

需要掌握的字词 Words and Expressions You Need to Grasp

label ['leɪbl] v. 把……贴标签	position [pə'zɪʃən] n. 职位
staff [stɑːf] n. 职员	by heart 牢记
carry out 执行	strict [strɪkt] adj. 严格的

练习 Exercises

根据所学内容填空。(Fill in the blanks according to what you have learnt.)

| label | by heart | carried out | positions |

① Make the duties of all the _____ clear.

② All the rules must be _____ strictly.

③ Clean up the unnecessary things, divide goods into groups and _____ them.

④ Make sure all staffs learn the rules in the kitchen _____ in various forms of trainings.

Part II Vocational Module
——(Some Useful English Dialogues on Cookery)
第二部分　职业模块
——烹饪英语常用对话

◎ **项目四**
点餐
Taking Orders

◎ **项目五**
菜肴特色介绍
Introducing Specialties

◎ **项目六**
菜肴制作过程介绍
Introducing the Cooking Ways

◎ **项目七**
厨房介绍
Introducing the Kitchen

◎ **项目八**
面试英语（烹饪）
The English Expressions Used for a Cook Job Interview

◎ **项目九**
厨师岗位
The Cook on the Post

项目四　点餐
Taking Orders

一、散餐；按菜单点菜
A La Carte

（W——Waiter　G——Guest）

W: Good evening, madam and sir. A table for two?

G: Yes, please. Could we have a table by the window?

W: I'm sorry, it's already reserved. Would you like this table?

G: OK, we'll take this one.

W: OK. Would you like to order now, madam?

G: I'd like to try some Sichuan dishes. What do you recommend?

W: Sichuan cuisine is well-known for its hot and spicy food. Today our restaurant serves a lot of Sichuan dishes, such as Kung Pao Chicken, Yu-Shiang Shredded Pork, Mapo Tofu and so on.

G: That sounds good. We'll have them.

W: Would you like a salad? We have potato salad, tomato salad and some fruit salads.

G: A potato salad, please.

W: And what would you like to drink?

G: I prefer dry wine.

W: Just a moment, please.

基础词汇 Basic Vocabulary

reserved [ri'zə:vd] *adj.* 预定的，保留的

spicy ['spaisi] *adj.* 辛辣的

serve [sə:v] *vt. & vi.* （为……）服务；提供，端上

prefer [pri'fə:] *vt.* 更喜欢

recommend [ˌrekə'mend] *vt.* 推荐，介绍

restaurant ['restərənt] *n.* 饭店，餐馆

salad ['sæləd] *n.* 沙拉，凉拌菜

重点短语和句型 Key Phrases and Sentences

A table for...　一桌……人的位置

Could we...　我们是否能……

Would you like...　你要……吗

and so on　等等

Would you like to order now?　您现在要点餐了吗？

What do you recommend?　你们有什么可以推荐的吗？

What would you like to drink?　你想要喝什么？

What would you like to eat?　你想要吃什么？

练习 Exercises

（1）If you are attending a job interview, here are some questions appearing in the test paper. Let's see whether you can pass the test or not.

周氏小厨餐厅招聘一名服务生，招聘考试内容如下。

① 来自英国的Mr. Bean向你点了一份fruit salad，请你从下列三幅图中挑出他点的那份沙拉并打钩。

（　　）　　　　　　（　　）　　　　　　（　　）

② 请将下列英文称呼，填在正确的图片下方。

　　　　　guest　　　　　　　　　　waiter

（　　）　　　　　　（　　）

③ 以下三个菜为今日的特价菜，请将这三个菜的英文名称写在菜单上，方便Mr. Bean挑选。

（2）以下是你在点单过程中，与Mr. Bean之间的对话，但顺序被打乱了，请你将它们整理配对，并在（　　）里填入相应字母。（Put the dialogue in order.）

（　　）1. A table for two?　　　　　　　　a）A fruit salad, please.

（　　）2. Would you like to order now, sir?　　b）Yes, please.

（　　）3. Would you like a salad?　　　　　c）I prefer dry wine.

（　　）4. What would you like to drink?　　　d）I'd like to try some Sichuan dishes.

（3）第（2）题中，Mr. Bean 正带着他的两个好朋友来到周氏小厨餐厅吃饭，而你正是负责点单的服务生，请你结合下面的图片与第（2）题中所给的提示，根据课文内容编一组小对话。（Make out a dialogue.）

今日菜单

二、自助餐
Buffet

（W——Waiter G——Guest）

W: Good afternoon, madam. Would you like buffet or a la carte?

G: Buffet. How much for one person?

W: The price of our buffet is 158 yuan for one person before 6:00 pm. And the price is 198 yuan for one in the evening. We'll give a 50% discount for the old（age＞60）and kids under 1.4 meter high.

G: OK. Thank you.

W: There are many kinds of dishes in our restaurant. You can have a try.

G: Thanks a lot.

基础词汇 Basic Vocabulary

buffet [ˈbʌfit] n. 自助餐

discount [ˈdiskaunt] n. 数目，折扣

kid [kid] n. 小孩；年轻人

price [prais] n. 价格，价钱；代价

meter [ˈmi:tə] n.〈美〉米；计，表，仪表

重点短语和句型 Key Phrases and Sentences

have a try   试一试

many kinds of   许多种的

How much...   ……多少钱

How much for one person?   一位多少钱？（在自助餐中常常这样提问）

give a（50%）discount for（the old）   给（老人）（50%）的折扣

Would you like（buffet）or（a la carte）？   您是想要（自助餐）还是（点菜）？

练习 Exercises

（1）今天Mr. Bean一个人来到周氏小厨餐厅用餐，尽管是一个人，但他还是希望可以吃得花样多些，但是每份菜的量不必太多。你会建议他使用哪种用餐方式？在你选择的图片下方打√，并写入相关单词。（Which type of dining should

(   )_____

(   )_____

Mr. Bean choose？）

（2）这是 Mr.Bean 与服务员之间的对话，但是问答之间的顺序有误，请你帮着将正确的序号写在（　　）里。（Put the dialogue in order.）

（　　）Would you like buffet or a la carte?

（　　）How much for one person?

（　　）What about the discount for the olds?

a）The price of our buffet is 158 yuan for one person.

b）Buffet.

c）We'll give a 50% discount.

（3）假设你是餐厅老板，请你设计一个优惠活动。在下面广告牌中写出你的优惠活动，并且结合课文内容，为你的新优惠活动编一组对话。（Design a poster and make out a dialogue.）

三、酒水
Beverage

（W——Waiter　G——Guest）

W: What would you like to drink, sir?

G: Give me a Whisky, please.

W: Straight up or on the rocks, sir?
G: Straight up. Ice will spoil the taste.
W: Enjoy your drink. Have a pleasant evening.

基础词汇 Basic Vocabulary

beverage [ˈbevərɪdʒ] *n.* 饮料

straight [streit] *adj.* 直的；*adv.* 直接地

spoil [spɔil] *vt.* 溺爱；糟蹋；掠夺

Whisky [ˈwiski] *n.* 威士忌酒

rock [rɔk] *n.* 岩石（这里指冰块）

taste [teist] *n.* 味道；品味

重点短语和句型 Key Phrases and Sentences

straight up   净饮

on the rocks   加冰块

What would you like...   你想要什么……

Enjoy your drink (meal...).   好好享用。

Have a pleasant (evening).   （夜晚）愉快。

练习 Exercises

（1）如果你是调酒师，你会将冰块放入哪杯酒中？请你用箭头符号表示冰块的去向。（Where to put the ice?）

（2）According to the dialogue, which one does the guest like better, straight up or on the rocks? Why? What about you?

（3）Mr.Bean和朋友一起来到了一家酒吧聚会，他在服务台点了一杯酒。而你正好是酒吧的调酒师，请你展开想象，设计一组对话。（Make out a dialogue.）

四、项目四总练习
Exercises for Project IV

Translate the following sentences into English.

（1）我想要个靠窗的桌子。

（2）我更喜欢喝干红。

（3）我们为年龄大于60周岁的老人和身高在1.4米以下的儿童提供百分之五十的折扣。

（4）抱歉，这张桌子已经被预订了。

（5）多少钱一位？（提示：自助餐中价格的提问）

项目五　菜肴特色介绍
Introducing Specialties

一、川菜
Sichuan Cuisine

（W——Waiter　G——Guest）

W: Good evening, sir. We cater to both Chinese and western taste. Which would you prefer, Chinese or western?

G: We'll have Chinese food for a change today. Frankly speaking, we want to try some spicy and hot dishes. Could you recommend us something special?

W: I would suggest you try some Sichuan dishes. Sichuan cuisine is famous for its hot, spicy food. It emphasizes the use of seasonings, so no two dishes ever taste alike. I would recommend Kung Pao Chicken, Yu-Shiang Shredded Pork, Mapo Tofu and Crucian Carp with Chili Bean Sauce.

G: Are all the Sichuan dishes hot?

W: No. Sichuan food is not always hot. Apart from hot food, we have sweet and sour food that many foreign friends like, light taste food and multi-flavoured food.

G: Good. Your recommendations. I want to try the sweet and sour dish.

W: Sweet and sour pork chops.

G: OK. We'll take it. And Yu-Shiang Shredded Pork, Mapo Tofu and Crucian Carp with Chili Bean Sauce.

W: OK.

基础词汇　Basic Vocabulary

cater ['keitə] vt. & vi. 提供饮食及服务　　　　taste [teist] n. 滋味，味道

frankly ['fræŋkli:] adv. 直率地（说），坦诚地　　cuisine [kwi'zi:n] n. 烹饪，烹调法

emphasize ['emfəsaiz] vt. 强调；加强语气；重读　seasoning ['si:zəniŋ] n. 调味品；佐料

light ['lait] *adj.* 轻的（这里指"清淡的"）
multi-flavoured ['mʌlti] ['fleivəd] *adj.* 多种风味的
recommendation [ˌrekəmen'deiʃən] *n.* 推荐；推荐信；建议

重点短语和句型 Key Phrases and Sentences
cater to　投合，迎合（在这里是"为……提供饮食服务"的意思）
Frankly speaking...　坦白说……
be famous for...　因……而得名，著名
apart from　除……之外
I want to...　我想……
Which would you prefer, (Chinese)or (western)?　你更喜欢哪个,（中式的）还是（西式的）？
We'll have Chinese food for a change today.　我们今天想换个口味。
Could you recommend us something special?　你能向我们推荐一些有特色的菜肴吗？
Your recommendations.　听你的（建议）。

练习 Exercises
知己知彼，方能百战百胜。作为一位川菜馆的老板，你了解自己的餐厅吗，你了解自己的顾客吗？（Suppose you are a boss of a Sichuan cuisine restaurant. Can you answer the questions below? ）

（1）What taste does the restaurant cater to?
___

（2）Which taste do the guests prefer, Chinese or western?
___

（3）What kind of dishes do the guests want to try?
___

（4）What is the Sichuan cuisine famous for?
___

（5）What does the waiter recommend for the guests?
___

（6）Are all the Sichuan dishes hot?
___

趣味阅读 Interesting Reading

川菜简介

川菜以成都、重庆两个地方菜为代表，选料讲究、规格划一、层次分明、鲜明协调。川菜作为我国八大菜系之一，在我国烹饪史上占有重要地位，它取材广泛、调味多变、菜式多样、口味清鲜、醇浓并重，以善用麻辣著称，并以其别具一格的烹调方法和浓郁的地方风味，融汇了东南西北各方的特点，博采众家之长，善于吸收、善于创新，享誉中外。

香辣蟹

（1）川菜的派系

①上河帮（蓉派，以成都和乐山菜为主）：其特点是小吃，亲民为主，比较清淡，传统菜品较多。蓉派川菜讲求用料精细准确，严格以传统经典菜谱为准，其味温和，绵香悠长，通常颇具典故。其著名菜品有麻婆豆腐、回锅肉、宫保鸡丁、盐烧白、粉蒸肉、夫妻肺片、蚂蚁上树、灯影牛肉、蒜泥白肉、樟茶鸭子、白油豆腐、鱼香肉丝、泉水豆花、盐煎肉、干煸鳝片、东坡墨鱼、清蒸江团等。

②下河帮（渝派，以重庆和达州菜为主）：其特点是家常菜，亲民，比较麻辣，多创新。渝派川菜大方粗犷，以花样翻新迅速、用料大胆、不拘泥于材料著称，俗称江湖菜。大多起源于市民家庭厨房或路边小店，并逐渐在市民中流传。渝派川菜近几年来在全国范围内大受欢迎，不少的川菜馆主要菜品均为渝派川菜。其代表作有酸菜鱼、毛血旺、口水鸡、干菜炖烧系列（多以干豇豆为主）；水煮肉片和水煮鱼为代表的水煮系列；辣子鸡、辣子田螺和辣子肥肠为代表的辣子系列；泉水鸡、烧鸡公、芋儿鸡和啤酒鸭为代表的干烧系列；泡椒鸡杂、泡椒鱿鱼和泡椒兔为代表的泡椒系列；干锅排骨和香辣虾为代表的干锅系列等。风靡海内外的麻辣火锅（或称毛肚火锅、火锅）发源于重庆，因为其内涵已超出川菜的范围，通常被认为是一个独立的膳食体系而不被视作川菜的组成部分。

③小河帮（盐帮菜，以自贡和内江为主）：其特点是大气，怪异，高端（其原因是盐商）。

一般认为蓉派川菜是传统川菜，渝派川菜是新式川菜。以做回锅肉为例，蓉派做法中材料必为三线肉（五花肉上半部分）、青蒜苗、郫县豆瓣酱以及甜面酱，缺一不可；而渝派做法则不然，各种带皮猪肉均可使用，青蒜苗也可用其他蔬菜代替，甜面酱用蔗糖代替。而具体烩制手法两派基本相似。不同之处在于蓉派沿袭传统，渝派推陈出新（虽然未必较传统做法更加美味）。

（2）其他资料

四川各地小吃通常也被看作是川菜的组成部分。主要有担担面、川北凉粉、麻辣小面、酸辣粉、叶儿粑、酸辣豆花等以及用创始人姓氏命名的赖汤圆、龙抄手、钟水饺、吴抄手等。

二、鲁菜
Shandong Cuisine

（W——Waiter　G——Guest）

W: Good evening, madam. Welcome to Qilu Restaurant.

G: Good evening.

W: Come this way, please.

G: Thank you.

W: Have a seat, please. Here is the menu. Are you ready to order, madam?

G: Yes. But there are so many dishes on the menu. It's difficult to decide. Would you please recommend some?

W: Sure. Our restaurant serves typical Shandong style dishes. Shandong cuisine is the most famous cuisine in northern China. There are a lot of typical Shandong dishes. I would recommend the Braised Sea Cucumber with Scallion, Yellow River Carp in Sweet and Sour Sauce.

G: Sounds really good. We'll take them. Thank you.

W: With pleasure.

基础词汇 Basic Vocabulary

typical ['tipikəl] *adj*. 典型的；特有的

cucumber ['kju:kʌmbə] *n*. 黄瓜，胡瓜

sour ['sauə] *adj*. 有酸味的，酸的

scallion ['skæljən] *n*. 青葱，冬葱，大葱

重点短语和句型 Key Phrases and Sentences

Welcome to...　欢迎来……

Come this way, please.　请走这边。

Would you please recommend some?　你可以推荐一些吗？（注意类似句型的积累）

Have a seat, please.　请坐。

Are you ready to order?　你准备好点餐了吗？

With pleasure.　很愿意效劳。

练习 Exercises

（1）请你用英文为以下两个菜命名。（Name these two dishes in English.）

（2）排序。（Put the dialogue in order.）

(　) Good evening, madam. Welcome to Qilu Restaurant.

(　) Have a seat, please. Here is the menu. Are you ready to order, madam?

(　) Sure. I would recommend the...

(　) Sounds really good. We'll take them. Thank you.

(　) With pleasure.

(　) Good evening.

(　) Yes. But there are so many dishes on the menu. Would you please recommend some?

趣味阅读 Interesting Reading

鲁菜简介

（1）历史与流派

鲁菜，又称山东菜，发端于春秋战国时的齐国和鲁国（今山东省），形成于秦汉。宋代后，鲁菜就成为"北食"的代表，是我国八大菜系之一。鲁菜是我国覆盖面较广的地方风味菜系，遍及京津塘及东北三省。

鲁菜主要由沿海的胶东菜（以海鲜为主）和内陆的济南菜以及自成体系的孔府菜组成。

鲁菜讲究调味纯正，口味偏于咸鲜，具有鲜、嫩、香、脆的特色。十分讲究清汤和奶汤的调制，清汤色清而鲜；奶汤色白而醇。

德州扒鸡　　　　　　　　爆双脆　　　　　　　　糖醋鲤鱼

（2）主要特点

庖厨烹技全面，巧于用料，注重调味，适应面广。其中尤以"爆、炒、烧、焖"等最具特色。烧有红烧、白烧，著名的"九转大肠"是烧菜的代表；"焖"是山东独有的烹调方法，其主料要事先用调料腌渍入叶或夹入馅心，再沾粉或挂糊。两面焖煎至金黄色，放入调料或清汤，以慢火焖尽汤汁，使之浸入主料，增加鲜味。山东广为流传的锅焖豆腐、锅焖菠菜等，都是久为人们所乐道的传统名菜。

九转大肠　　　　　　　　百花大虾　　　　　　　　蟹黄海参

鲁菜还精于制汤。汤有"清汤""奶汤"之别。经过长期实践，清汤现已演变为用肥鸡、肥鸭、猪肘子为主料，经沸煮、微煮、清哨，使汤清澈见底，味道鲜美。奶汤则呈乳白色。用"清汤"和"奶汤"制作的数十种菜，多被列为高级宴席的珍馐美味。

烹制海鲜有独到之处。对海珍品和小海味的烹制堪称一绝。在山东，无论是参、贝，还是鳞、虾、蟹，经当地厨师妙手烹制，都可成为精彩鲜美的佳肴。仅胶东沿海生长的比目鱼（当地俗称"偏口鱼"），运用多种刀工处理和不同技法，可烹制成数十道美味佳肴，其色、香、味、形各具特色，百般变化于一鱼之中。以小海鲜烹制的"油爆双花""红烧海螺""炸蛎黄"以及用海珍品制作的"扒原壳鲍鱼""绣球干贝"等，都是独具特色的海鲜珍品。

鲁菜善于以葱香调味，在菜肴烹制过程中，不论是爆、炒、烧、熘，还是烹调汤汁，都以葱丝（或葱末）爆锅，就是蒸、扒、炸、烤等菜，也借助葱香提味，如"烤鸭""烤乳猪""锅烧肘子""炸脂盖"等，均以葱段为佐料。

三、淮扬菜
Huaiyang Cuisine

(A and B are friends, they are traveling in Yangzhou)

A: Yangzhou is very beautiful. I like the city.

B: Me too. And the food here is also very famous.

A: Really? Can you tell me something about Yangzhou food?

B: OK. Huaiyang cuisine is one of the most famous Chinese cuisines in China. It occupies an important position in Chinese cuisine. And it is represented mainly by Yangzhou style, Suxi style, Jinling style and Xuhai style, each having a distinct feature. Yangzhou food is light while Suxi food is sweet. Being crisp and tender is the characteristic of Jinling food, whereas the flavour of Xuhai food is salty and fresh.

A: Sounds inviting! What shall we have in the evening?

B: In my opinion, we can have a local try. Dried Bean Curd Shreds in Chicken Soup, Steamed Minced Pork and Crab Balls, Sweet and Sour Mandarin Fish are very famous. And Steamed Bread with Three Different Dices tastes good.

A: That sounds delicious. Let's go.

基础词汇 Basic Vocabulary

occupy [ˈɔkjupai] vt. 占据，占领；居住
represent [ˌrepriˈzent] vt. 代表；表现
feature [ˈfiːtʃə] n. 特色，特征
inviting [inˈvaitiŋ] adj. 诱人的；有魅力的
taste [teist] vi. 尝起来

position [pəˈziʃən] n. 方位；地位，身份
mainly [ˈmeinli] adv. 主要地，大体上
whereas [wɛərˈæz] conj. 然而；反之
local [ˈləukəl] adj. 当地的；地方性的

重点短语和句型 Key Phrases and Sentences

one of ...　其中之一
In my opinion ...　在我看来……

练习 Exercises

（1）地方与口味。（Tell something about Yangzhou food according to the dialogue.）

Yangzhou food is_____.

Suxi food is_____.

Xuhai food is＿＿＿＿＿＿＿＿＿＿＿＿＿＿＿＿＿.

Jinling food is＿＿＿＿＿＿＿＿＿＿＿＿＿＿＿＿＿.

（2）Mr.Bean和他的朋友想吃淮扬菜，结果服务员上菜时，送错了一盘菜，请你在菜单上找出这盘菜，并在菜名后打上×。（Find out the wrong dish.）

Menu

Dish	
Dried Bean Curd Shreds in Chicken Soup	( )
Steamed Minced Pork and Crab Balls	( )
Yellow River Carp in Sweet and Sour Sauce	( )
Sweet and Sour Mandarin Fish	( )
Steamed Bread with Three Different Dices	( )

趣味阅读 Interesting Reading

淮扬名菜

中国菜系素来就有八大风味，四大菜系的分类。淮扬菜与鲁菜、川菜、粤菜并称为中国四大菜系。淮扬菜，始于春秋，兴于隋唐，盛于明清，素有"东南第一佳味，天下之至美"之美誉。许多标志性事件的宴会中，都是淮扬菜唱主角。

淮扬菜是长江中下游（扬子江）、淮河中下游的代表风味。扬州是淮扬菜的中心和发源地。

淮扬菜十分讲究刀工，刀工比较精细，尤以瓜雕享誉四方。菜品形态精致，滋味醇和；在烹饪上则善用火候，讲究火功，擅长炖、焖、煨、焐、蒸、烧、炒；原料多以水产为主，注重鲜活，口味平和，清鲜而略带甜味。著名菜肴有清炖蟹粉狮子头、大煮干丝、三套鸭、水晶肴肉、松鼠鳜鱼、梁溪脆鳝等。其菜品细致精美、格调高雅。

 博里羊肉（贵妃羊肉）  淮安茶馓（金线缠臂）  平桥豆腐（西施豆腐）  文楼涨蛋

| 软兜长鱼（嫦娥善舞） | 人间第一鲜——文楼蟹黄汤包 | 淮饺（炒米馄饨） | 虾米扒蒲菜（红玉列兵） |

淮扬八大名宴

汉代以来，扬州宴席闻名全国。明清时期，扬州宴席发展到一个高潮。康熙《扬州府志》载："涉江以北，宴会珍错之盛，扬州为最。"

| 满汉全席 | 板桥宴 | 红楼宴 | 鉴真素宴 |

| 开国第一宴 | 乾隆宴 | 三头宴 | 少游宴 |

四、项目五总练习
Exercises for Project V

Translate the following sentences into English.

（1）我们店提供中餐和西餐服务。

_____

（2）您想吃中餐还是西餐？

_____

（3）今天我们想换个口味吃中餐。
　　_____

（4）你能给我们推荐些特色菜吗？
　　_____

（5）除了麻辣味的菜肴，我们这还有许多外国朋友都喜欢吃的糖醋类菜肴和复合味菜肴。
　　_____

（6）我们餐厅主营粤菜，包括潮州菜、汕头菜和客家菜。
　　_____

（7）淮扬菜在中国烹饪界中占据很重要的地位。
　　_____

（8）扬州菜清淡，苏锡菜偏甜，金陵菜脆嫩，徐海菜咸鲜。
　　_____

项目六　菜肴制作过程介绍
Introducing the Cooking Ways

一、中餐
Chinese Food

1. 热菜 Hot Dishes

（W——Waiter　G——Guest）

W: Your Beijing Roast Duck, sir.

G: Let's have a try... Perfect! So tender and crisp! I've never tasted anything like it before. Could you tell me how to make it?

W: OK. The preparation is a bit complicated. Let me give you a brief introduction. First, duck must be split open, dressed, scalded and dried. When roasting, it's better to use fruit tree branches as firewood to lend more flavor to the duck. Besides, a steady temperature must be maintained in the oven and ducks must be roasted so that they are evenly roasted.

G: How long does it take to roast the duck?

W: About 50 minutes. And when the skin turns crisp and golden brown, the duck is done.

G: Interesting!

基础词汇 Basic Vocabulary

roast [rəust] vt.& vi. 烤；烘；焙
perfect ['pə:fikt] adj. 完美的，理想的
crisp [krisp] adj. 脆的，鲜脆的
complicated ['kɔmplikeitid] adj. 复杂的
bit [bit] n. 少量，少许
firewood ['faiəwud] n. 木柴
flavor = flavour['fleivə] n. 味；味道
temperature ['tempəritʃə] n. 温度
oven ['ʌvən] n. 烤箱，炉

duck [dʌk] n. 鸭，母鸭；鸭肉
tender ['tendə] adj. 温柔的；柔软的
preparation [ˌprepə'reiʃn] n. 预备；准备
brief [bri:f] adj. 简洁的，简短的
scald [skɔ:ld] vt. 烫
lend [lend] vt. 增加，增添
steady ['stedi] adj. 稳定的，坚定的
maintain [mein'tein] vt. 维持；继续
even ['i:vən] adj. 均匀的

重点短语和句型 Key Phrases and Sentences

have a try　试一试

I've never tasted anything like it before.　我之前从来都没有尝过像这样的味道。

Let me give you a brief introduction.　让我来为您大致介绍一下。

练习 Exercises

（1）Mr.Bean来到北京，点了当地最有名的北京烤鸭，他特别喜欢这道菜，于是请教了你这些问题，请你用英文告诉他。（Questions about Beijing Roast Duck.）

① What is this in English?
It is _____.

② How does it taste? _____.

③ How to make it?
First, duck must be _____, _____, _____ and _____.
Secondly, _____.

④ How long does it take to roast the duck? _____

（2）Mr.Bean喜欢收集各地名菜，他会将听到的菜谱制作方法等记录在自己的本子上。假设你是Mr.Bean，请你结合课文内容以及第（1）题练习的内容，以北京烤鸭为主线，写一篇简短的饮食日记。（Food diary.）

Food diary

_____
_____
_____
_____
_____
_____
_____

2. 凉菜 Cold Dishes

(W——Waiter   G——Guest   C——Chef)

W: Is everything to your satisfaction?

G: Yes, thank you. Your asparagus salad tasted very good. Could I make it at home?

W: I'm sorry I don't understand. I'll ask the chef to help you.

A few minutes later …

C: Good evening, madam and sir. May I help you?

G: The asparagus salad is very delicious. We like it very much. I just want to know how could I make the dish at home?

C: Thank you, madam. I'm glad to hear that. It's so easy. First, cut 1 pound asparagus diagonally, bring 4 cups water to boil in saucepan. Then drop in asparagus, boil 1 minute, drain, rinse with cold water. Mix next four ingredients (2 tablespoons light soy sauce, 2 tablespoons sesame oil, 1/4 teaspoon sugar and chopped garlic) in a bowl.

G: Is that all?

C: Almost. The last step is to pour the dressing over the asparagus. It's OK. By the way, the dressing may be kept in covered jar in the refrigerator for about a week.

G: Wow, it sounds nice. I may try it someday.

C: Hope you enjoy staying with us.

G: Thank you!

基础词汇 Basic Vocabulary

diagonally [daiˈægənəli] *adv.* 对角地；斜对地

boil [bɔil] *vt. & vi.* （使）沸腾；用开水煮，在沸水中煮

saucepan [ˈsɔːspæn] *n.* 深平底锅

drain [drein] *vt. & vi.* （使）流干，（使）逐渐流走

rinse [rins] *vi.* 冲洗掉；漂净

tablespoon [ˈteɪbəlˌspuːn] *n.* 大汤匙；一大汤匙的量

teaspoon [ˈtiːˌspuːn] *n.* 茶匙；一茶匙的量

chop [tʃɔp] *vt. & vi.* 砍，伐，劈

dressing [ˈdresiŋ] *n.* 穿衣；加工；调味品

bowl [bəul] *n.* 碗，钵，盘；一碗之量

pour [pɔː] *vt. & vi.* 涌出；倾；倒

jar [dʒɑː] *n.* 罐子，广口瓶；（啤酒）杯

重点短语和句型 Key Phrases and Sentences

by the way　顺便说说；顺便问一下

Is everything to your satisfaction?　您对一切都还满意吗？

"to one's satisfaction"　意思是"使某人满意"。

We like it very much.　我们很喜欢它。

I'm glad to hear that.　我很高兴听到这个。

Hope you enjoy staying with us.　希望您在这里过得愉快。

练习 Exercises

（1）你知道这些职业的英文名称吗？请你写在横线上。（What is their occupation?）

_____　　_____　　_____

（2）下面图片中有一份冷菜和一份热菜，你能正确地将它们分辨，并写出英文表达方式吗？（Name the hot dish and cold dish below.）

_____   _____

（3）请你帮助Mr.Bean列出制作芦笋沙拉的过程。（How to make the asparagus salad?）

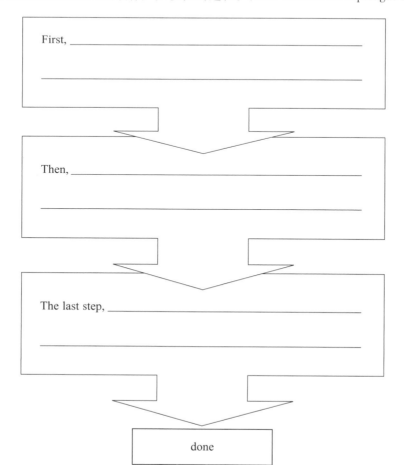

二、项目六总练习
Exercises for Project VI

Translate the following sentences into English.

（1）我从来没有吃过这么美味的菜肴。
_____

（2）烤鸭的制作过程有些复杂。
_____

（3）首先，把鸭子开膛、燂毛、烫皮、晾干。
_____

（4）烤的时候最好用果木燃料，这样鸭子的味道会更香。
_____

（5）对不起，这个我不太懂。我马上联系厨师来帮助您。
_____

（6）首先，把一磅的芦笋斜切，再在锅里倒入4杯水烧沸。
_____

（7）然后放入芦笋，焯水，控水，过凉。
_____

项目七　厨房介绍
Introducing the Kitchen

一、中式厨房
Chinese Kitchen

（W——Western Cook　C——Chinese Cook）

W: It's a chopping block, right?

C: Yes. We cut meat and other things on it. Look this, it's a walk-in refrigerator. It's large. We usually put food in it to keep it fresh. And we also put the meat in it to thaw.

W: What's this?

C: It's a casserole. We Chinese cook usually make soup in it. It can maintain the temperature and keep the flavor very well. And we also stew some food that is difficult to be cooked in it.

W: Oh, I see. I've heard you Chinese cook usually steam the pastries in the bamboo steamer. Where is it? I want to have a look.

C: This way, please.

Modern Chinese Kitchen

Old Chinese Kitchen

基础词汇 Basic Vocabulary

chopping [ˈtʃɔpiŋ] *n.* 截断
chopping block 砧板；剁肉板
refrigerator [riˈfridʒəreitə] *n.* 冰箱
casserole [ˈkæsəˌrəul] *n.* 焙盘；砂锅
pastry [ˈpeistri] *n.* 糕点

block [blɔk] *n.* 块；街区
walk-in [ˈwɔːkˌin] *n.* 可供人走进之物
thaw [θɔː] *vi.* 融解；变暖和
temperature [ˈtempəritʃə] *n.* 温度，气温
bamboo [bæmˈbuː] *n.* 竹，竹竿

重点短语和句型 Key Phrases and Sentences

have a look   看一看
This way, please.   这边请。

练习 Exercises

请根据所给提示，将该物的图片序号写在括号里，并为该物体注上英文的表达方式。
（Answer the questions below according to the pictures.）

  a   b   c   d

（1）We cut meat and other things on it. （　　）_____

（2）It's large. We usually put food in it to keep it fresh. And we also put the meat in it to thaw. （　　）_____

（3）We Chinese cook usually make soup in it. It can maintain the temperature and keep the flavor very well. （　　）_____

（4）Chinese cook usually steam the pastries in it. （　　）_____

二、厨房设备与工具
Kitchen Utensil and Cooking Equipment

1. 盛器 Dishware

（M——Mrs. Green　T——Tom）

M: Tom, can you give me a hand with something in the kitchen?

T: Yes, mum. What do you want me to do?

M: First of all, I need you to clean the plates and dishes.

T: OK. Where should I put them?

M: Put them in the cupboard, please. Then come to help me with the soup pots and the wine glasses. It's really hard to clean them.

T: Oh, these pots are so greasy! And the glasses are terrible, too.

M: Take it easy. Use the detergent and the dish cloth, please.

T: All right. Let me deal with them, mum.

M: Thank you, dear!

基础词汇 Basic Vocabulary

greasy ['gri:zi] *adj.* 油腻的

terrible ['terəbl] *adj.* 很糟糕；可怕的

重点短语和句型 Key Phrases and Sentences

deal with   处理

first of all   首先

Give somebody a hand with...   帮助某人做某事，相当于 help sb.（to）do sth.

It's hard / easy / dangerous / kind to do something.
做某事困难、简单、危险、友好等。

Take it easy.   别紧张，别急。

练习 Exercises

（1）填空。（Fill in the blanks.）

① These pots are so_____（油腻的）! And the glasses are_____（可怕的）.

② It's really_____（困难的）to clean them.

③ _____（首先），I need you to clean the plates and dishes.

（2）翻译句子。（Translate these following sentences.）

① 你能帮我在厨房干点活吗？

② 你想让我干什么呢？

③ 帮我洗一下汤盆和酒杯。

④ 让我来处理它们。

2．加热用具 Cookware

（M——Mary D——Daniel）

M: Hi, Daniel. I need to buy lots of things to furnish my new kitchen. Can you give me some advice?

D: Sure, I'm glad to help you. What kinds of things do you need right now?

M: Well, I think I need the cookware immediately because I have invited my parents to have dinner tomorrow.

D: OK. Do you have any special requirement?

M: Not really. Actually I have no idea about it. You know I'm not good at shopping.

D: Well, according to your situation, I suggest you to buy a pressure cooker, a gas range, and an automatic cooker. Besides, you can buy a pan as well as an oven. These things are all necessary.

M: Good idea! Thank you for your suggestions.

D: And you can buy other things later, such as the boiler. It's good for boiling food.

M: Many thanks! I'll follow your advice.

D: My pleasure!

基础词汇 Basic Vocabulary

immediately [iˈmi:diətli] *adv.* 立刻地

requirement [riˈkwaiəmənt] *n.* 要求

situation [ˌsitjuˈeiʃən] *n.* 情况

necessary [ˈnesisəri] *adj.* 必要的；必然的 *n.*[pl.]必需品

重点短语和句型 Key Phrases and Sentences

lots of　很多　相当于 a lot of, 后面可接可数名词复数或不可数名词

I'm glad to...　我很高兴……

right now　现在，目前

not really　事实上没有，事实上不是

have no idea about sth.　对某物没有想法，没有主意

according to　根据

suggest sb. to do sth.　建议某人做某事

such as　比如

It's good for doing sth.　用来做某事不错，适合用作于……

Can you give me some advice?　你能给我一些建议吗？

I'll follow your advice.　我会听从你的建议的。

My pleasure!　别客气；我很荣幸！

练习 Exercises

（1）将图片放入厨房中正确的位置。(Match Task.)

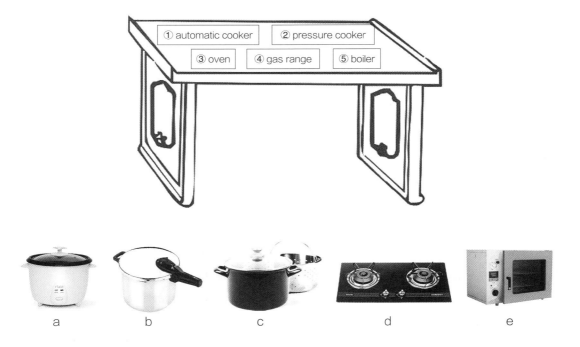

（2）根据对话内容，回答问题。(Answer the questions according to the dialogue.)

① What kinds of things does Mary need right now?

　　*She needs*_____

② Why does Mary need the cookware immediately?

　　*Because she has*_____

③ Does Mary have any special requirement?

　　_____

④ What is Daniel's suggestion?

　　*He suggests Mary to*_____

⑤ What is the boiler good for?

　　_____

3. 辅助设备 Auxiliary Equipment

（M——Mary　A——Amy）

M: Come in, Amy! Let me show you my new kitchen!

A: Wow! Your kitchen is really nice. Is this the new refrigerator?

M: Yes, you are eagle-eyed. I bought it last week. You see, I had a separate refrigerator and freezer, but these have both combined into one.

A: That's cool! Is this automatic dryer new, too?

M: Yes, it saves time. I needn't to dry dishes any more.

A: What about this one? Is this an ice breaker?

M: You are right. I love iced tea. It's wonderful, especially in summer. Oh, I've got an electric juicer here. Look! What do you think of it?

A: Wow, it's great! I love juice. I'm thinking of buying one, too.

M: How about drinking a glass of juice? I made some this morning.

A: Great! Thank you.

基础词汇 Basic Vocabulary

eagle-eyed ['i:gl'aid] adj. 有眼力的
freezer ['fri:zə] n. 冰箱；冷冻库
automatic [ˌɔ:tə'mætik] adj. 自动的
especially [i'speʃəli] adv. 尤其是
wonderful ['wʌndəful] adj. 极好的，惊人的

separate ['sepəreit] adj. 单独的；分开的
combine [kəm'bain] v. 结合 n. 集团
breaker ['breikə] n. 打破者
ice breaker 破冰设备

重点短语和句型 Key Phrases and Sentences

save time　节省时间

not... any more　不再，相当于not... any longer和no longer

What do you think of it?　你觉得它怎么样？

think of doing sth.　考虑做某事

How about doing sth.　表示提建议，（去）……怎么样

练习 Exercises

（1）根据课文完成对话。(Complete the dialogue according to the text.)

A: Your kitchen is really nice. Is this the new_____?

B: Yes, you are_____. I bought it last week.

A: What about this one? Is this_____?

B: Yes, it_____time. I needn't to dry dishes_____.

A: _____about this one? Is this an_____?

B: You are right. I love iced tea. It's_____, _____in summer.

B: What do you think of_____?

A: Wow, it's_____! I love juice. I'm thinking of_____one, too.

（2）把下列设备名称与图片正确配对。（Match Task.）

① refrigerator   ② ice breaker   ③ automatic dryer   ④ electric juicer

a　　　　　　　　b　　　　　　　　c　　　　　　　　d

4．厨房设备 Kitchen Utensil

（C——Chef　V——Visitor）

C: Welcome to our kitchen. Now please allow me to give some introduction.

V: Thank you, please.

C: Firstly, this way, please. This is our freezing compartment.

V: Oh, it's very spacious and tidy. There is so much seafood here.

C: Yes, you're right. We keep them here for freshment.

V: What's this? Is it a machine?

C: Yes, it's an ice machine used for making ice. And this is the ice breaker. We use it to break the ice into small ones.

C: This way, please. This is the most advanced oil smoker exhausting equipment.

V: Oh, it's a very technical one, isn't it?

C: You're right. This one is very effective. Now let me show you another utensil.

V: ...

C: ...

基础词汇 Basic Vocabulary

compartment [kəmˈpɑːtmənt] n. [建] 隔间；区划

tidy [ˈtaidi] adj. 整齐的；相当大的

technical [ˈteknikəl] adj. 技术的，专业的

oil smoker exhausting equipment n. 抽排油烟设备

spacious [ˈspeiʃəs] adj. 宽敞的

advanced [ədˈvɑːnst] adv. 先进的

effective [iˈfektiv] adj. 有效的

重点短语和句型 Key Phrases and Sentences

allow sb. to do sth.　允许某人做某事

be used for doing sth.　用来做某事

show sb. sth.　展示某物给某人，相当于 show sth. to sb.

练习 Exercises

（1）回答问题。（Answer questions.）

① We keep seafood for fresh. What is it?_____

② We use it for making ice. What is it?_____

③ We use it for breaking the ice. What is it?_____

④ We use it for absorbing（吸）the smoke. What is it?_____

（2）根据要求介绍厨房。（Introduce the kitchen according to the requirement.）

请至少选择以下设备中的三种进行厨房介绍。

refrigerator	ice breaker	automatic dryer	electric juicer
automatic cooker	oven	pressure cooker	gas range
ice machine	boiler	freezing compartment	
oil smoker exhausting equipment			

_____

_____

_____

_____

三、项目七总练习
Exercises for Project Ⅶ

Translate the following into English.

（1）你能来厨房帮我做点家务活吗？

_____

（2）这些东西真的很难洗。

_____

（3）我需要购买很多东西来装修我的新厨房。
　　_____

（4）我想我急需一些加热用具，因为我邀请了我的父母明天来吃饭。
　　_____

（5）根据你的情况，我建议你买一个高压锅，一个煤气灶和一个电饭煲。
　　_____

（6）非常感谢，我会接受你的建议。
　　_____

（7）来一杯果汁如何？
　　_____

（8）现在让我来给你介绍一下吧。
　　_____

（9）喔，它很宽敞，也很整齐。
　　_____

（10）这是最先进的抽排油烟设备。
　　_____

项目八　面试英语（烹饪）
The English Expressions Used for a Cook Job Interview

一、自我介绍
Self-introduction

（A——the Interviewer　　B——the Interviewee）

A: Good morning.

B: Good morning, sir.

A: Would you please introduce yourself briefly?

B: Yes. My name is Lin Hai. I'm from Zhejiang province. I'm an active boy and I like sports. Cooking and basketball are my favorites.

A: OK, good. What's your strongpoint to be as a cuisine major?

B: I think I'm patient and friendly. And this kind of characteristic is a strongpoint for a cuisine major. Also, I'm a hard-working student and like to learn from others.

A: Very good! Then do you have any weakness?

B: Yes, frankly speaking, sometimes I'm not very good at planning. Now I have realized it and I am trying to improve this weakness.

A: OK, this is good. I think you can overcome it. Thank you for your self-introduction.

B: Thank you very much, sir.

基础词汇 Basic Vocabulary

introduce [ˌintrəˈdju:s] *v.* 介绍

strongpoint [ˈstrɔŋˈpɔint] *n.* 优点

characteristic [ˌkæriktəˈristik] *n.* 性格

frankly [ˈfræŋkli] *adv.* 坦诚地

overcome [ˌəuvəˈkʌm] *v.* 克服

briefly [briːfli] *adv.* 简单地

patient [ˈpeiʃənt] *adj.* 有耐心的

weakness [ˈwiːknis] *n.* 弱点

realize [ˈriəlaiz] *v.* 意识到

重点短语和句型 Key Phrases and Sentences

My name is（Lin Hai）. 我的名字叫（林海）。

I'm from（Zhejiang province）. 我来自（浙江省）。或者可以说 I come from（Zhejiang province）.

learn from others　向别人学习

frankly speaking...　坦白说……

on time　准时

Would you please...?　你能……吗？（客气的说法）

Would you please introduce yourself briefly?　你能简单介绍一下自己吗？

improve the weakness　改善缺点

What's your strongpoint to be as a cuisine major?　作为一名烹饪学生你的优势是什么？

Thank you for（your self-introduction）. 谢谢（你的自我介绍）。

练习 Exercises

根据对话写一份你自己的英文简历。（Write an English resume about yourself.）

Resume

Name: _____ Age: _____

From: _____

Characteristic: _____

Strongpoint: _____

Weaknesses: _____

二、面试对话
Interview

(A —— the Interviewer   B —— the Interviewee)

A: Good morning.

B: Good morning, sir.

A: OK, firstly would you explain why you choose cuisine as your major?

B: Yes. I choose cuisine as my major because I love cooking. I really enjoy creating a meal from various ingredients and watching my family or friends enjoy it. It gives me a real sense of satisfaction.

A: OK, just now you said you enjoy cooking, right? Then, what do you think of cooking?

B: I think cooking needs patience, carefulness and practice. Impatience can not make a good cook.

A: Good. And what is your career purpose?

B: I want to be a successful chef.

A: Oh, that sounds great. But as we know, it's not easy to become a qualified chef. What will you do to achieve your career goal?

B: Firstly, I will study as hard as I can. Secondly, I will try to practice more and communicate with others. Thirdly, I will pay attention to my comprehensive quality and improve my abilities.

A: OK, good, that's all. Thank you.

B: Thank you very much, sir.

基础词汇 Basic Vocabulary

ingredient [inˈgri:diənt] n. 原料
patience [ˈpeiʃəns] n. 耐性，耐心
successful [səkˈsesfəl] adj. 成功的
qualified [ˈkwɔliˌfaid] adj. 有资格的，能胜任的

satisfaction [ˌsætisˈfækʃən] n. 满足
impatience [imˈpeiʃəns] n. 不耐烦，急躁
chef [ʃef] n. 厨师，大师傅
achieve [əˈtʃi:v] vt. 达到，实现

重点短语和句型 Key Phrases and Sentences

enjoy doing sth.　喜欢、享受做某事
What do you think of doing sth.?　你觉得……怎么样？
firstly, secondly, thirdly　第一，第二，第三；用于陈述并列的几个观点
as we know　众所周知
achieve one's career goal　实现某人的事业目标
communicate with...　与……交流
pay attention to sth.　关注某事某物
comprehensive quality　综合素质
Would you explain why you choose cuisine as your major?　你可以解释一下为什么你会选择烹饪作为你的专业吗？
It gives me a real sense of satisfaction.　它能给我一种真正的成就感。
What is your career purpose?　你的事业目标是什么？

练习 Exercises

（1）回答下列面试问题。（Answer the interview questions below.）

面试考题

① why do you choose cuisine as your major?

_____

② what do you think of cooking?

_____

③ what is your career purpose?

_____

④ what will you do to achieve your career goal?

_____

(2)根据课文内容和所回答的问题,找一个搭档,分别扮演面试者和经理人,看看谁演的最好!写下关键的词句。(Role play.)

三、项目八总练习
Exercises for Project VIII

Translate the following sentences into English.

(1)你能简单介绍一下自己吗?

(2)我是一个活泼的男孩,我喜欢运动。烹饪和足球是我的最爱。

(3)作为一名烹饪专业学生你的优势是什么?

(4)现在我意识到了我的弱点,我正在努力改进。

(5)我相信你能克服的。

(6)谢谢你的自我介绍。

(7)我是一个勤勉的学生,喜欢向他人学习。

(8)你可以解释一下为什么你会选择烹饪作为你的专业吗?

(9)它能给我一种真正的成就感。

(10)你的事业目标是什么?

（11）我会尽可能地努力学习。
_____

（12）我想成为一名成功的主厨。
_____

项目九　厨师岗位
The Cook on the Post

一、从学徒工开始做起
Start from a commis

（A——the Interviewer　B——the Chef）

A: Hi! Tom. I heard that you have been working in many restaurants. Now you become a chef. Can you talk about the process of becoming a chef? What do you do to get a job in a kitchen?

B: Well, there're lots of different ways, but the most common way is to start as a commis first, which means you go to the kitchen and work at the bottom level doing the basic chopping —— boring jobs —— for maybe two or three years. You do get paid by your employer but it's a really small wage.

A: So what about you? Did you go to cooking school?

B: Yeah, I learnt many basic skills at school and started my internship as a commis in a restaurant at the last year of my study. It really helped me a lot to learn much experience.

A: It is the hard-working that helps you improve a lot. How much do you think you learn by watching others and how much do you learn on your own, using your own creativity?

B: I think both are really important. Actually, I learned a lot from my mother. When I was a kid, I used to watch her cooking all the time, and it wasn't until I grew up that I realized how much I understood about cooking just from seeing what she did in the kitchen, but also talking about how to do things with your colleagues, I think, is really important.

A: I see. Thank you for the interview and I hope I can have a chance to enjoy a bite by your cooking.

B: It would be my pleasure.

基础词汇 Basic Vocabulary

process [ˈprəʊses] *n.* 过程

employer [ɪmˈplɔɪə(r)] *n.* 雇主

wage [weɪdʒ] *n.* 工资

internship [ˈɪntɜːnʃɪp] *n.* 实习

actually [ˈæktʃuəli] *adv.* 事实上

bottom [ˈbɒtəm] *n.* 底部

level [ˈlevl] *n.* 水平

skill [skɪl] *n.* 技能；本领

creativity [ˌkriːeɪˈtɪvəti] *n.* 创造力

colleague [ˈkɒliːɡ] *n.* 同事

重点短语和句型 Key Phrases and Sentences

talk about…　谈论……

at the bottom level　在底层；在底端

get paid　得到报酬，领工资

What about you?　你呢？

It really helped me a lot.　对我的帮助真的很大。

I learned a lot from…　我从……学习到很多。

not… until…　直到……才……

not only… but also…　不仅……而且……

grow up　长大

enjoy a bite　品尝

It would be my pleasure.　那将是我的荣幸。

练习 Exercises

（1）看下面的图片，试着找出哪些工作可以属于学徒工。再根据所学内容，试着写下这些工作是由什么职位负责的。(Look at all the pictures below and try to find out what kinds of jobs can belong to a commis. Then, try to write down the position which is in charge of each job according to what you have learnt.)

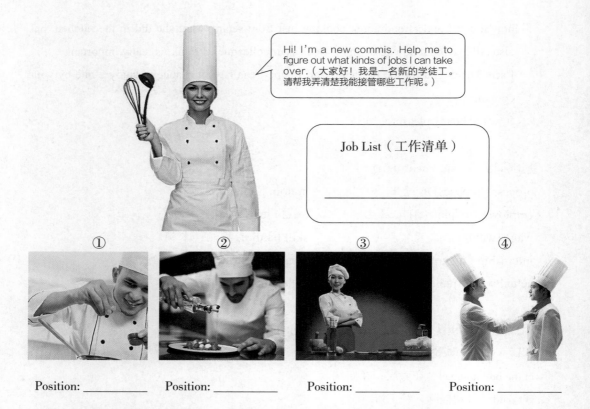

Position: _____   Position: _____   Position: _____   Position: _____

（2）试着思考更多学徒工可以做的工作，然后用英语说给搭档听。（Try to think about more jobs a commis can do and talk to your partner in English.）

二、协作在厨房

Cooperate with others in the kitchen

（J——Jack, the Commis   T——Tom, the Chef   S——the Sous-chef）

S: Jack, I would like to introduce you to Tom. He is our Executive Chef. Tom, this is Jack, a commis.

J: Nice to meet you, Tom. I'm a new-comer here. I hope you can help me.

T: Nice to meet you, too, Jack. Believe me, everyone starts from new. I think you'll adapt quickly. If you get any problem, just tell me.

J: That's so nice of you. So anything I can do now?

T: Ok. Please clean this pot. I need to use it 5 minutes later.

J: No problem, Tom. I will do it right now.

T: Thank you! Hurry up!

J: Here is the pot, Tom. I washed it twice.

T: Not this one!

J: I'm sorry! I must take the wrong one. Here it is. It's the one you need.

T: Thank you! Good job, Jack.

J: My pleasure.

基础词汇 Basic Vocabulary

introduce [ˌɪntrəˈdju:s] *v.* 介绍　　executive [ɪgˈzekjətɪv] *n.* 行政领导

new-comer *n.* 新来的人　　adapt [əˈdæpt] *v.* 适应

pot [pɒt] *n.* 壶，盆　　pleasure [ˈpleʒə(r)] *n.* 快乐

重点短语和句型 Key Phrases and Sentences

I would like to introduce you to Tom.　我想把你介绍给汤姆。

This is...　这位是……（常用语介绍他人）

I'm a new-comer here.　我是新来的；我初来乍到。

Everyone starts from new.　每个人都是从新手做起的。

That's so nice of you.　你真是太好了。

No problem!　没问题！　　right now　立刻，马上

Hurry up!　赶快！　　Good job!　干得不错！

My pleasure.　我的荣幸。

练习 Exercises

试着根据我们刚学的对话，跟搭档编写对话。情况如下。（Try to make out a dialogue with your partner, according to the dialogue we've just learnt. The situation is like this.）

山姆初来厨房上班。他是一位学徒工。他要和鱼菜厨师乔治一起工作。行政总厨汤姆要把山姆介绍给乔治。当山姆和乔治一起工作时，他被要求处理一条三文鱼，但他拿错了鱼。

Sam is a new-comer to the kitchen. He is a commis. He is going to work with the Fish Cook George. The Executive Chef, Tom will introduce Sam to George. When working with George, Sam is asked to deal with a salmon but he takes the wrong one.

I'm Tom, an Executive Chef. I'm going to introduce Sam to George.

I'm George, a Fish Cook. There is a new-comer in our department.

I'm a commis and a new-comer. I take the wrong salmon. What should I do?

三、厨房准则
Dos and don'ts in the kitchen

(J—— Jack, the Commis　　T—— Tom, the Chef)

T: Jack, put this pork into the refrigerator.

J: Okay, but how can I deal with it?

T: Store it in a clean and dry container.

Jack puts the pork into a right container.

J: Is it okay, Tom?

T: Good job, Jack.

J: Can I put it into refrigerator now?

T: Not yet, it should be labeled with the contents and date.

J: Okay! I see.

T: What about the soup, Jack? Is it ready?

J: I'm not sure, but according to the timer, yes.

T: Taste it, Jack. You'll know.

J: So how should I taste the soup?

T: Spoon from the soup into this spoon. Then pour the soup into this small spoon. Now taste it!

J: Wow. It is very delicious! I think it's ready.

T: Well done, Jack! Do remember never taste the soup over an open stockpot.

J: I'll bear in mind, Tom.

基础词汇 Basic Vocabulary

refrigerator [rɪˈfrɪdʒəreɪtə(r)] n. 冰箱
content [kənˈtent] n. 内容
timer [ˈtaɪmə(r)] n. 计时器
store [stɔː(r)] v. 储藏

container [kənˈteɪnə(r)] n. 容器
date [deɪt] n. 日期
pour [pɔː(r)] v. 倒
stockpot [ˈstɒkpɒt] n. 汤锅

重点短语和句型 Key Phrases and Sentences

deal with... 处理……
pour... into... 把……倒入……
bear in mind 记住；牢记

label... with... 把……贴上……（标签）
well done 做得好

练习 Exercises

（1）根据对话内容，帮助杰克列出他在厨房能做和不能做的事。（Help Jack list something he can do and something he can't in the kitchen according to the dialogue.）

Dos	Don'ts
_____	_____
_____	_____
_____	_____

（2）你能列出更多厨房准则吗？跟你的搭档讨论一下。（Can you list more dos and don'ts in the kitchen? Discuss with your partner.）

Dos	Don'ts
_____	_____
_____	_____
_____	_____

四、厨房安全
Safety in the kitchen

(A——the Interviewer   B——the Chef)

A: Tom, we're talking about working in the kitchen. When I would help out in the kitchen, I was always afraid of the big knives and the fires and the burns and stuff, so can you talk a little about safety in the kitchen?

B: Yeah, that's really important actually. First there's the uniform. You have to cover as much of your skin as you can to avoid burns and if you have a special chef jacket, it must be all cotton, so if you get something hot on it, it will still be safe, and it can be quickly taken off.

A: What should we do to make the burn go away when we are burned?

B: You have to be careful not to touch the burn or break it. You should of course immediately put in under cold water and then afterwards use vitamin E oil and that is really good, actually. It makes the skin heal really well.

A: What about cuts?

B: Learn how to chop to keep all your fingers out away from the knife and remember you always have the knife in contact with your hand so you don't need to look at it when you cut and you don't cut yourself.

A: That's pretty impressive. Thanks for the safety tips.

基础词汇 Basic Vocabulary

safety ['seɪfti] *n.* 安全
cotton ['kɒtn] *adj.* 棉制的
afterwards ['ɑːftəwədz] *adv.* 后来
impressive [ɪm'presɪv] *adj.* 令人印象深刻的

avoid [ə'vɔɪd] *v.* 避免
immediately [ɪ'miːdiətli] *adv.* 立刻
heal [hiːl] *n.* 痊愈
tip [tɪp] *n.* 小建议

重点短语和句型 Key Phrases and Sentences

help out… 帮助摆脱困境
take off 起飞；脱下
of course 一定；当然
That's pretty impressive. 这真是太让人印象深刻了；这真是太好了。
Thanks for… 为……而感谢

be afraid of 害怕
go away 走开；离去
in contact with 接触；与……有联系

练习 Exercises

（1）根据对话内容，给出小建议。(Give the tips according to the dialogue.)

How to avoid getting burnt in the kitchen? 如何避免在厨房里烧伤？
_____
_____
_____

How to avoid getting cut in the kitchen? 如何避免刀伤？
_____
_____
_____

（2）根据下面给出的小建议造对话。(Make a dialogue according to the tips given below.)

- 用干毛巾绕着烫锅柄。
  Use a dry towel to circle around the hot panhandle.
- 记住即使是微潮的毛巾预热会产生水蒸气，而水蒸气的灼伤是最严重的。
  Remember even a damp towel will create steam when it becomes hot and steam burns are the worst of all.

towel 毛巾　　　　　circle 围绕　　　panhandle 锅柄
damp 潮湿的　　　　steam 水蒸气

五、项目九总练习
Exercises for Project IX

Translate the following into English.
（1）你能谈一谈成为主厨的历程吗？

（2）你确实从雇主处领到工资但少得可怜。
_____

（3）它确实对我学习经验帮助很大。
_____

（4）我希望我可以有机会尝一尝你的手艺。
_____

（5）我想你会很快适应的。
_____

（6）你真是太好了。
_____

（7）它应该标上食品和日期的标签。
_____

（8）一定要记住不要在开盖的汤锅上方尝汤。
_____

（9）谢谢这些安全小建议。
_____

（10）我会牢牢记住。
_____

Part III Extended Module
第三部分 拓展模块

◎ **项目十**
西式餐点简介
Brief Introduction of Western Cuisine

◎ **项目十一**
西餐主题口语对话
Dialogues with the Topics of Western Cuisine

项目十　西式餐点简介
Brief Introduction of Western Cuisine

一、西餐基本烹饪法
Basic Cooking Methods of Western Cuisine

教学指引 Teaching Guideline

在接下来的学习中，我们将向大家介绍西式餐点的一些基本知识，主要包括西餐基本烹饪法、西餐食品原料、西餐用餐礼仪等。在本节的教学内容中，我们着重介绍西餐基本烹饪方法的相关语言知识。掌握烹饪方法的表达法，可以帮助学生更好地掌握西餐知识，更广泛地阅读相关资料。

在本节的教学中，教师可以根据学生的学习情况，结合其专业知识，适当地拓展学习内容，并根据学生的兴趣和学习需求，介绍一定的课外知识。在教学过程中，建议教师灵活发挥，充分结合学生实际需求，创造丰富有趣的语言课堂。

知识背景介绍 The Introduction of Knowledge Background

各种食物无论用什么方法完成烹调，都需要不同的热源传送才能完成，例如燃气（燃气炉、烤箱）、电力（电磁炉、电烤箱、微波炉）、蒸汽（蒸笼）和油脂（油炸）等。了解热的传送方式和速度，有利于更好地完成烹饪。热的传递方式主要有三种：直接传热、间接传热或热对流传热和辐射传热。

烹饪方式则可分为湿热法和干热法两大类。其中，湿热法主要有以下几种。

（1）低温水煮　适合煮柔嫩的鲜鱼和荷包蛋。温度设定在71℃~82℃。

（2）过水、过油　适合去除骨头的血水与杂质和咸火腿与培根的盐分，以及烫青菜、马铃薯去除有害的酶，避免与空气接触而产生变色。温度设定在85℃~96℃。

（3）沸煮　适合煮面类、水饺、带壳的蛋等。温度设定在100℃。注意水量要够，避免放入食物后，温度降低太多使沸腾时间增长。

（4）蒸　适合保持外形完整及减少水溶性维生素流失的方法。在100℃沸滚的时候放入食品原料，以热气传导将食物蒸熟。

（5）焖　先以干热法使食物表面变色，再以少量液体焖制完成。

以上是几种主要的湿热法，接下来我们来了解一下干热法的常见方式。

（1）烤及烘烤　由干热的空气围绕食物完成烹调。

（2）炙烤　以红外线辐射热源，由上往下以高温快速完成烹饪。

（3）炭烤　热源由下往上加热完成烹饪。

（4）翻炒　在锅内加入少量油脂，快速加热完成烹饪。

（5）煎　在锅里放入适量油脂，以中火加热，两面煎成变色即完成烹饪。

（6）油炸　将食物放入175℃~190℃之内的热油中，加热完成烹饪。

学生学习 Student Learning

导入 Lead-in

◎ 思考题 Questions
① 思考：基本的西餐知识有哪些？你了解哪些西餐烹饪法？
② 讨论：根据你的生活经历和专业知识，谈谈你对基本西餐烹饪法的看法。

基础烹饪方法 Words-Learning: Basic Cooking Methods

西餐烹饪方法主要有湿热法和干热法两大类。不同的烹饪方法有不同的烹饪效果，适用于不同的食材和不同的加工需求。接下来，我们向大家介绍几种常用的烹饪方法及其适用的菜肴。

（1）湿热法

① 过水、过油 blanching

Blanched Broccoli 过水西蓝花　　　　　Blanched French Bean 过水四季豆

Blanched Chicken Bone 过水鸡骨　　　　Blanched French Fries 过油炸薯条

> **提示 Tips**
>
> 食物原料过水分为两种方式：一是少量食物原料放入大量的滚水中，短时间烫热，如菠菜叶、西蓝花、四季豆等；二是大量的食品原料放入冷水中不盖锅盖慢火煮滚以去除多余的杂质或盐分，如咸火腿、培根、鸡骨等。

② 低温水煮 poaching

Poached Egg 荷包蛋　　　　　　　　Poached Fish Fillet 温煮鱼片

Poached Chicken Breast 水煮鸡胸

> **提示 Tips**
>
> 　　低温煮可防止食物外表逐渐变干。海鲜、家禽类经常用鱼汤或高汤进行低温水煮，而肉类、烟熏类、蛋类等则用水作低温处理。

③ 沸煮 boiling

Boiled Egg 煮蛋　　　　　　　　　　Boiled Chicken Bone Stock 煮鸡高汤

④ 蒸 steaming

Steamed Whole Sea Bass 蒸鲈鱼　　　Steamed Cabbage Roll 蒸包心菜卷

◎ 相关词汇 Related Words

Convection Steamer 高压式电气蒸箱

Convection Oven 对流烤箱

⑤ 焖 braising

Osso Bucco 意式烩小牛膝

Chicken Cacciatore 意大利烩鸡

意大利烩鸡

（2）干热法

① 烤 roast

Roast Stuffed Turkey 烤瓤火鸡

Roast Beef 烤牛肉

Roast Lamb Chops in Cheese and Red Wine Sauce 烤羊排配奶酪和红酒汁

② 烘烤 bake

Baked Lobster with Garlic and Butter 巴黎黄油烤龙虾

Baked Chicken Breast Stuffed with Mushroom and Cheese 烤鸡胸酿奶酪蘑菇馅

③ 炭烤 grill

Grilled Stuffed Chicken Rolls 烧瓤春鸡卷

Grilled Rib-Eye Steak 扒肉眼牛排

④ 翻炒 sauté

Pan-Fried Duck Breast with Sautéed Potatoes and Truffles

秘制鸭胸配黑菌炒马铃薯

⑤ 油炸 deep-fry

Deep Fried Breaded Tiger Prawn 炸明虾　　　　Flour Fried Chicken Basket 竹篮炸鸡

Plain Fry French Fries 清炸薯条　　　　　　　Batter Fried Onion 面糊炸洋葱圈

⑥ 煎 pan-fry

Pan Fried Fillet of Beef 煎菲力牛排　　　　　Pan Fried Fillet of Cod 煎鳕鱼排

⑦ 炙烤 broil

扒肉眼牛排　　　　　煎鳕鱼排　　　　　面糊炸洋葱圈　　　　巴黎黄油烤龙虾

需要掌握的字词 Words and Expressions You Need to Grasp

blanch [blɑ:ntʃ] v. 过水；漂白　　　　　　broccoli [ˈbrɔkəli] n. 西蓝花，花椰菜

poach [pəutʃ] v. 水煮　　　　　　　　　　bass [bæs] n. 鲈鱼

fillet [ˈfilit] n. 束发带；鱼肉片　　　　　　convection [kənˈvekʃən] n.（热的）对流

braise [breiz] vt. 蒸，炖，焖　　　　　　　roast [rəust] v. 烤，炙，烘；n. 烤肉

broil [brɔil] vt. 烤　　　　　　　　　　　　prawn [prɔ:n] n. 对虾，明虾

练习 Exercises

西餐的基本烹饪法有哪两大类？分别有哪几种烹饪方式？请结合实例进行简述。

二、西餐食品原料
Raw Food Material of Western Cuisine

教学指引 Teaching Guideline

本节课的学习内容是常见的西餐食品原料。西餐食品原料和中餐食品原料既有相同的部分，又有不同的部分。因此，在学习西餐食品原料的时候，一方面可以借助前面已学的中餐食品原料知识，通过已知语言知识来实现新语言知识的掌握；另一方面，要关注两者之间的区别，把握要领，掌握精髓。

西餐常用的食品原料主要包括海鲜、畜肉、禽类、鸡蛋、奶制品、蔬菜、水果和粮食。其中，每类食品原料又有更细的分类及加工品。各式各样的食品原料容易造成学生在学习上的困扰，尤其是对相关词汇的记忆。因此，在教学过程中，可以从简到易，先让学生掌握较为熟悉的语言知识，再通过操练，扩展到较难层次的内容。教师可以营造活跃的课堂氛围，并通过不同形式的词汇教学法，引导学生攻克这一难点。

知识背景介绍 The Introduction of Knowledge Background

西餐原料知识是以西餐行业中经常使用的烹饪原料为研究对象，对其外形、结构等各方面知识进行综合阐述的，所以它是西餐烹饪专业的一门基础学科。如此基础的课程，在烹饪英语中必然要涉及，只有有了这样的知识沉淀，才会有后续更好的学习和提升。

学生学习 Student Learning

导入 Lead-in

◎ 思考题 Questions

① 还记得前面学过的中餐食品原料吗？它们有哪些分类，分别具有什么样的特点？

② 讨论：你觉得中西方饮食习惯上有哪些不同？西餐食品原料与中餐食品原料有哪些异同点？

食品原料 Words-Learning: Raw Food Material

接下来，我们按类别对常见的西餐食品原料进行简单的介绍。

（1）奶制品 milk products

milk 牛奶	evaporated milk / condensed milk 炼乳
dry milk 奶粉	buttermilk / sour milk 酸奶
yogurt 酸奶酪	cheese 奶酪
cream 奶油	ice cream 冰淇淋
butter 黄油	

提示 Tips

牛奶（milk）根据不同的成分特征和加工形式，又可以分为很多种，例如常见的全脂牛奶和低脂牛奶。

试一试 Have a try

根据下面提供的中文和英文单词，将英文单词与之中文意思进行连线。

whole milk　　　　　　冷冻果汁牛奶
low-fat milk　　　　　　全脂牛奶
skim milk　　　　　　　低脂牛奶
ice milk　　　　　　　　撇取牛奶
sherbet　　　　　　　　冷冻牛奶

（2）畜肉 meat
beef 牛肉　　　　　　　veal 小牛肉
lamb 羊肉　　　　　　　pork 猪肉

提示 Tips

西餐使用量最大的畜肉是牛肉，其次是羊肉、小牛肉和猪肉。

（3）家禽 poultry
chicken 鸡肉　　　　　　turkey 火鸡
duck 鸭　　　　　　　　goose 鹅
guinea fowl 珍珠鸡　　　　pigeon 鸽子

（4）水产品 fish and shellfish
freshwater fishes 淡水鱼

saltwater fishes 海水鱼

shellfish 贝壳水产品

水产品又可以分为甲壳水产品（crustacean）和软体水产品（mollusk），常见的水产品有：龙虾（lobster）、海蟹（crab）、鲜贝（scallop）、鱿鱼（squid）等。

（5）植物原料 plant material

vegetable 蔬菜　　　　　　　　fruit 水果

potatoes and starches 马铃薯和淀粉类原料

需要掌握的字词 Words and Expressions You Need to Grasp

evaporated [i'væpəreitid] *adj.* 浓缩的，脱水的，蒸发干燥的

condensed [kən'denst] *adj.* 浓缩的　　　cheese [tʃi(:)z] *n.* 乳酪，干酪

yogurt ['jəugə:t] *n.* 酸奶，酵母乳　　　veal [vi:l] *n.* 小牛肉

sherbet ['ʃə:bət] *n.* 冰果子露，冷冻牛奶　　turkey ['tə:ki] *n.* 火鸡，火鸡肉

poultry ['pəultri] *n.* 家禽，禽肉　　　shellfish ['ʃelfiʃ] *n.* 贝，甲壳类动物

pigeon ['pidʒin] *n.* 鸽子

练习 Exercises

（1）根据图片中的食材，写出其英文名。（Name the food according to the pictures.）

_____　_____　_____　_____

（2）还记得课文中提到的水产品吗？试着将下列词汇进行中英连线。（Match Task.）

freshwater fishes　　　　　　　　　贝壳水产品

saltwater fishes　　　　　　　　　　软体水产品

shellfish　　　　　　　　　　　　　淡水鱼

crustacean　　　　　　　　　　　　甲壳水产品

mollusk　　　　　　　　　　　　　海水鱼

三、西餐烘焙简介
The Brief Introduction of Baking

教学指引 Teaching Guideline

本节课的学习内容是西餐烘焙中常见的设备、原料和成品。同样，会有些与中餐相重叠的词汇，在让学生借力学习的同时也要注意区分。

烘焙技术是西点制作中主要的应用技术。在欧美国家，烘焙食品在人民的饮食中占据着十分重要的地位，面包是欧美许多国家人民的主食，其工业化、自动化的发展，对减轻广大人民的家务劳动、使饮食方便化、合理化以及节约能源、解放生产力起了巨大的推动作用。就连祖祖辈辈以大米为主食的日本，面包类的消费也是惊人的。因此在学习西餐英语中，烘焙英语的相关知识不可或缺。

知识背景介绍 The Introduction of Knowledge Background

烘焙，又称为烘烤、焙烤，是指在物料燃点之下通过干热的方式使物料脱水变干变硬的过程。烘焙是面包、蛋糕类产品制作不可缺少的步骤，通过烘焙后淀粉产生糊化、蛋白质变性等一系列化学变化后，面包、蛋糕达到熟化的目的。烘焙也能使食物的口感发生变化。简而言之，烘焙食品是以粮油、糖、蛋等为原料基础，添加适量辅料，并通过和面、成形、焙烤等工序制成的口味多样、营养丰富的食品。

学生学习 Student Learning

导入 Lead-in

◎ 思考题 Questions

（1）回顾一下在西点课堂上学习到的烘焙常见原料有哪些？哪些已经会用英语表达了？

（2）讨论：西方的烘焙技术和我们中式面点制作的技术有何异同？

烘焙简介 Words-Learning: The Brief Introduction of Baking

1. 常用设备 Common Facilities

baking pan	烤盘	dough mixer	搅拌机
piping bag	裱花带	fermenting box	醒发箱
piping tube	裱花嘴	oven	烤炉
spatula	抹刀	scale	秤
thermometer		温度计	
measuring cup		量杯	
electric egg beater/mixer		电动打蛋器	

需要掌握的字词 Words and Expressions You Need to Grasp

baking [ˈbeɪkɪŋ] n. 烘焙
piping [ˈpaɪpɪŋ] v. 用管道输送
tube [tju:b] n. 管状物
spatula [ˈspætʃələ] n. 抹刀
oven [ˈʌvn] n. 烤箱
thermometer [θəˈmɒmɪtə(r)] n. 温度计
electric [ɪˈlektrɪk] adj. 电动的

dough [dəʊ] n. 生面团
mixer [ˈmɪksə(r)] n. 搅拌器
beater [ˈbi:tə(r)] n. 搅拌器
scale [skeɪl] n. 刻度；天平
ferment [fəˈment] v. 发酵
measure [ˈmeʒə(r)] v. 测量

练习 Exercises

根据图片，写出设备英文名。(Name all the equipment below in English.)

2. 常用原料 Common Raw Materials

Flour（面粉类）			
bread flour	面包粉	cake flour	蛋糕粉
rye flour	黑麦面粉	whole wheat flour	全麦粉
wheat bran	麦片	oatmeal	燕麦片
starch	淀粉	low-gluten flour	低筋面粉
high-gluten flour	高筋面粉		

Milk Products（奶制品类）			
butter	黄油	margarine	人造黄油
powdered milk	乳粉	cheese	奶酪
yoghurt	酸奶	sour cream	酸奶油

Sugar（糖类）			
brown sugar	红糖	cube sugar	方糖
icing sugar	糖粉	malt sugar	麦芽糖
granulated sugar	砂糖	lactose	乳糖
syrup	糖浆	honey	蜂蜜

Auxiliary Materials（辅料类）			
cocoa powder	可可粉	almond	杏仁
yeast	酵母	shredded coconut	椰丝
gelatin	鱼胶片	agar	琼脂

续表

Auxiliary Materials（辅料类）			
corn starch	玉米淀粉	bread improver	面包改良剂
baking soda	小苏打	chocolates	巧克力

需要掌握的字词 Words and Expressions You Need to Grasp

flour [ˈflaʊə(r)] n. 面粉　　　　　　　rye [raɪ] n. 黑麦
bran [bræn] n. 麸，糠　　　　　　　oatmeal [ˈəʊtmiːl] n. 燕麦片
starch [stɑːtʃ] n. 淀粉　　　　　　　gluten [ˈɡluːtn] n. 面筋
butter [ˈbʌtə(r)] n. 黄油　　　　　　margarine [ˌmɑːdʒəˈriːn] n. 人造黄油
powder [ˈpaʊdə(r)] n. 粉　　　　　　cube [kjuːb] n. 立方
malt [mɔːlt] n. 麦芽　　　　　　　　granulated [ˈɡrænjə,leɪtɪd] adj.（白糖）成砂状的
lactose [ˈlæktəʊs] n. 乳糖　　　　　　syrup [ˈsɪrəp] n. 糖浆
cocoa [ˈkəʊkəʊ] n. 可可粉　　　　　almond [ˈɑːmənd] n. 杏仁
yeast [jiːst] n. 酵母　　　　　　　　agar [ˈeɪɡɑː(r)] n. 琼脂
gelatin [ˈdʒelətɪn] n. 明胶　　　　　soda [ˈsəʊdə] n. 苏打

练习 Exercises

（1）根据图片中的食材，写出其英文名。（Name all the food materials below.）

_____　　_____　　_____　　_____

（2）还记得课文中提到的各种糖吗？试着将下列词汇进行中英连线。（Remember those kinds of sugar mentioned in the text? Try to match those sugar to their right Chinese meanings.）

brown sugar　　　　　　　糖粉

cube sugar　　　　　　　　砂糖

icing sugar　　　　　　　　麦芽糖

malt sugar　　　　　　　　方糖

granulated sugar　　　　　乳糖

lactose　　　　　　　　　　红糖

3. 成品 Products

cookie	曲奇饼干	tart	蛋挞
finger sponge	手指饼干	fruit pie	水果派
French bread	法式面包	Danish pastry	丹麦面包
sweet roll	甜卷面包	puff pastry	起酥；松饼
toast	吐司	donut	甜甜圈
pudding	布丁	mousse	慕斯
Tiramisu	提拉米苏	sponge cake	海绵蛋糕

需要掌握的字词 Words and Expressions You Need to Grasp

sponge [spʌndʒ] *n.* 松软布丁　　　　Danish [ˈdeɪnɪʃ] *adj.* 丹麦的

pastry [ˈpeɪstri] *n.* 面粉糕饼　　　　French [frentʃ] *adj.* 法国的

roll [rəʊl] *n.* 卷　　　　　　　　　　puff [pʌf] *n.* 泡芙

mousse [muːs] *n.* 慕斯　　　　　　　toast [təʊst] *n.* 吐司

pudding [ˈpʊdɪŋ] *n.* 布丁　　　　　　Tiramisu [ˌtɪrəməˈsuː] *n.* 提拉米苏

练习 Exercises

根据图片，写出食物英文名。(Name all the food below in English according to the pictures.)

四、西餐用餐礼仪
Dining Etiquette of Wsetern Cuisine

教学指引 Teaching Guideline

随着社会经济的发展和文化交流的全球性趋势，西餐在中国越来越受欢迎。各式西餐餐厅在各大城市闪现，并逐渐成为年轻人就餐的热门选择。在本节的学习中，我们将向大家介绍一些基本的西餐用餐礼仪知识，帮助大家在了解西方文化的同时，熟悉西餐用餐礼仪，从而更好地享受西餐。

西餐是西方国家各种西式餐点的统称，具有有别于中餐的显著特色。当然，不同的国家之间，其用餐礼仪也会有一些小小的差别。本节内容将以美国为例，选取一些典型话题，介绍用餐的基本知识，并在此基础上引导学生掌握相关语言知识。教师可以根据课堂实际情况和学生的学习需求，灵活运用课本，适当补充一些具有趣味性的课外知识，让学生更好地感受西方文明和西餐文化。本节的教学内容，也是对之前所学内容的一个总结。教师可以借助本节教学，融合学生已学知识，复习相关语言要点，实现学以致用、巩固知新的有效教学。

知识背景介绍 The Introduction of Knowledge Background

（1）餐具使用礼仪　吃西餐，必须注意餐桌上餐具的排列和置放位置，不可随意乱取乱拿。正规宴会上，每一道食物、菜肴即配一套相应的餐具（刀、叉、匙），并以上菜的先后顺序由外向内排列。进餐时，应先取左右两侧最外边的一套刀叉。每吃完一道菜，将刀叉合拢并排置于碟中，表示此道菜已用完，服务员便会主动上前撤去这套餐具。如尚未用完或暂时停顿，应将刀叉呈八字形左右分架或交叉摆在餐碟上，刀刃向内，意思是告诉服务员，我还没吃完，请不要把餐具拿走。

使用刀叉时，尽量不使其碰撞，以免发出大的声音，更不可挥动刀叉与别人讲话。

（2）进餐的顺序　一餐内容齐全的西菜一般有七八道，主要由以下几部分构成。

第一，饮料（果汁）、水果或冷盆（又称开胃菜）。目的是增进食欲。

第二，汤类（即头菜）。需用汤匙，此时一般上有黄油、面包。

第三，蔬菜、冷菜或鱼（也称副菜）。可使用垫盘两侧相应的刀叉。

第四，主菜（肉食或熟菜）。肉食主菜一般配有熟蔬菜，此时要用刀叉分切后放餐盘内取食。如有沙拉，需用沙拉匙、沙拉叉等餐具。

第五，餐后食物。一般为甜品（点心）、水果、冰淇淋等。最后为咖啡，喝咖啡应使用咖啡匙、长柄匙。

（3）面包等可用手取食　进餐时，除用刀、叉、匙取送食物外，有时还可用手取。如吃鸡、龙虾时，经主人示意，可以用手撕着吃。吃饼干、薯片或小粒水果，可以用手取食。面包则一律手取，注意取自己左手前面的，不可取错。取面包时，左手拿取，右手撕开，再把奶油涂上去，一小块一小块撕着吃。不可用面包蘸汤吃，也不可一整块咬着吃。

（4）用汤匙舀着喝汤　喝汤时，切不可以汤盘就口，必须用汤匙舀着喝。姿势是：左手扶着盘沿，右手用匙舀，不可端盘喝汤，不要发出吱吱的声响，也不可频率太快。如果汤太烫时，应待其自然降温后再喝。

（5）不可整块肉送嘴里咬　吃肉或鱼的时候，要特别小心。用叉按好后，慢慢用刀切，切好后用叉子进食，千万不可用叉子将其整个叉起来，送到嘴里去咬。这类菜盘里一般有些生菜，往往是用于点缀和增加食欲的，吃不吃由你，不要为了面子强吃下去。

（6）需要服务请侍者　餐桌上的佐料，通常已经备好，放在桌上。如果距离太远，可以请侍者或别人，不能自己站起来伸手去拿，这是很难看的。

（7）吃东西不要发出很大声响　吃西餐时相互交谈是很正常的现象，但切不可大声喧哗，放声大笑，也不可抽烟，尤其在吃东西时应细嚼慢咽，嘴里不要发出很大的声响，更不能把刀叉伸进嘴里。至于拿着刀叉做手势在别人面前挥舞，更是失礼和缺乏修养的行为。

（8）坐姿要端正　吃西餐还应注意坐姿。坐姿要正，身体要直，脊背不可紧靠椅背，一般坐于座椅的四分之三即可。不可伸腿，不能跷起二郎腿，也不要将胳臂肘放到桌面上。

（9）酒杯不斟满，喝酒不劝酒　饮酒时，不要把酒杯斟得太满，也不要和别人劝酒（这些都不同于中餐）。如刚吃完油腻食物，最好先擦一下嘴再去喝酒，免得让嘴上的油渍将杯子弄得油乎乎的。干杯时，即使不喝，也应将酒杯在嘴唇边碰一下，以示礼貌。

学生学习 Student Learning

导入 Lead-in

西餐用餐常分为两种：餐馆用餐和家庭聚餐。西方人重视事前预约，因此外出用餐往往要提前预订，尤其是知名餐馆，不提前预订通常需要等候多时。家庭聚餐则强调准时赴约，尤其要注意的是通常不提前到访。在学习过程中，我们要注意这些文化上的差异。

场景 1. Situation One（餐馆用餐）

Step one：Choosing a Restaurant 选择餐馆

Step two: Reserving a Table 预订餐位

Step three: Arriving at Restaurant 到达餐馆

Step four：Summoning a Waiter 召唤侍者

Step five：Ordering Food 点餐

Step six: Enjoying the Meal 用餐

Step seven: Paying the Bill 付款

◎ 对话 Dialogue

(H —— Host   C —— Customer)

H: This is the Gourmet kitchen. May I help you?

C: I'd like to make a reservation for Sunday afternoon.

H: What time?

C: How about 2:00 p.m.?

H: How many people?

C: Let me see, it'll be nine people, I think.

H: OK, let me have a look. Sorry, we don't have nine-person table at 2:00 p.m., but it is available at 3:00 p.m. Would you mind coming at 3:00 p.m.?

C: Well, that's quite late. How about 1:00 p.m.?

H: That's OK. We'll have a nine-person table at 1:00 p.m.

C: That will do. Thank you!

H: My pleasure!

背景知识介绍 Background knowledge

男女一起在餐馆用餐，通常由男方负责点菜（order）和付账（pay the check or bill）。许多家庭饭前要由家中一员致辞。

场景2. Situation Two（家庭聚餐）

◎ 餐位 Seating Arrangement

	Lady	Lady	Gentleman of honor	
Host	Lady of honor	Gentleman	Gentleman	Hostess

餐位就座表

背景知识介绍 Background Knowledge

餐位的安排大致如上图。原则上男主宾（gentleman of honor）坐在女主人（hostess）右边，女主宾（lady of honor）坐在男主人（host）右边，而且多半是男女相间而坐，夫妇不坐在一起，以免各自聊家常话而忽略与其他宾客间的交际。

◎ 餐具摆法 Tableware Arrangement

家庭或餐厅宴会时，餐具的种类和数量，因餐会的正式（formal）程度而定。越正式的餐会，刀叉盘碟摆得越多。本文所举的例子，适用于不是十分正式的宴会（多数家庭式宴会属于此类）。

① 叉子（forks）放在主菜盘（main plate）左侧，刀子（knives）、汤匙（spoons）摆在右侧。

② 刀叉和汤匙依使用的先后顺序排列。先用的放在离主菜盘最远的外侧，后用的放在离主菜盘近的内侧。假如主人决定先上主菜再上沙拉，就要把主菜叉子放在沙拉叉子的外侧。

③ 沙拉盘放在靠主菜盘的左边。美国人通常把主菜和沙拉一起送上桌来，而不像法国人一样，主菜吃完以后才上沙拉。

一般餐具摆设图（午宴、晚宴均适用）

① butter plate and knife 奶油碟子和奶油刀

② dessert spoon 甜点匙

③ glass 饮料杯

④ salad plate 沙拉盘

⑤ napkin 餐巾

⑥ main course fork 主菜叉子

⑦ salad fork 沙拉叉子

⑧ main plate 主菜盘

⑨ main course knife 主菜刀

⑩ soup spoon 汤匙

⑪ cup, saucer and teaspoon 茶（咖啡）杯、碟和茶匙

回忆一下，上述哪些表达方式是你学过的？

◎ 结论 Conclusion

记得要入境随俗（While in Rome, do as the Romans do.）。对于吃西餐的规矩有疑问时，留意您的男女主人，照着他们的样子做，一般就不会出错了。

背景知识介绍 Background Knowledge

美国人宴客，是由女主人（或男主人）先拿起餐具进食，客人才跟着动餐具。如果您不喜欢递过来的菜肴时，只要说"No, thank you."即可。咀嚼食物，一定要避免出声。用过的刀、叉，必须放回盘子里，不能放在餐桌巾上。吃完主菜，把刀和叉平行地斜放在主菜盘（main plate）上，是向主人或侍者表示可以把主菜盘拿走了。席间要轻声谈些轻松愉快的话题，尽量避免一声不响地闷着头吃饭。

◎ 对话 Dialogue

下面是一段模拟的席间对话。

Hostess: Would you like to have some more chicken?

Guest: No, thank you. The chicken is very delicious, but I'm just too full.

Host: But I hope you save room for dessert. Mary makes good pumpkin pies.

Guest: That sounds very tempting. But I hope we can wait a little while, if you don't mind.

Host: Of course. How about some coffee or tea now?

Guest: Tea, please. Thanks.

需要掌握的字词 Words and Expressions You Need to Grasp

reserve [ri'zə:v] vt. 保留；预订

available [ə'veiləbəl] adj. 可得到的

napkin ['næpkin] n. 餐巾

summon ['sʌmən] vt. 召唤，传唤

seating ['si:tiŋ] n. 座位；席位

attorney [ə'tə:ni] n. 律师

练习 Exercises

在西方国家，到餐厅用餐往往需要提前预订。根据你所学的知识，按照订餐用餐的先后顺序将下列词组进行排序，并在括号中注明它们的中文意思。（Make the following phrases in right order according to the reservation steps in western countries.）

Reserving a Table　　　　　Enjoying the Meal　　　　　Summoning a Waiter

Choosing a Restaurant　　　Ordering Food　　　　　　Paying the Bill

Arriving at Restaurant

（1）_____（　　　　）

（2）_____（　　　　）

（3）_____（   ）

（4）_____（   ）

（5）_____（   ）

（6）_____（   ）

（7）_____（   ）

项目十一　西餐主题口语对话
Dialogues with the Topics of Western Cuisine

一、点餐
Taking Orders

（W——Waiter　G——Guest）

W: May I take your order, madam?

G: May I have a look at the menu first?

W: Sure. Please take your time. I'll be back soon.

A few minutes later ...

G: Waiter, I'd like a steak.

W: OK. How would you like it prepared? Rare, medium or well-done?

G: Rare for me, please.

W: Fine. What would you like to go with your steak?

G: French fries and tomato salad, please.

W: Which soup do you prefer? We have mixed mushroom cream soup and seafood consommé.

G: Mixed mushroom cream soup, please. That's all.

W: So, you have ordered a mixed mushroom cream soup, a French fries and a tomato salad. A steak, rare. Is that right?

G: Yes.

基础词汇 Basic Vocabulary

steak [steik] *n*. 牛排
rare [rɛə] *adj*. 三分熟的
well-done [ˈwelˈdʌn] *adj*. 全熟的
mixed [mikst] *adj*. 混合的
seafood [ˈsiːˌfuːd] *n*. 海产食品，海鲜

prepare [priˈpɛə] *vt*. 准备
medium [ˈmiːdjəm] *adj*. 中等的，适中的
soup [suːp] *n*. 汤，羹
mushroom [ˈmʌʃrum] *n*. 蘑菇
consommé [kənˈsɔmei] *n*. [法语] 清炖肉汤

重点短语和句型 Key Phrases and Sentences

have a look at sth. 看一看某物
May I take your order? 我能为你点餐了吗？（服务员常在为客人点餐时使用）
Take your time. 慢慢来，别着急。
How would you like it prepared / cooked? 你的牛排要几分熟的？
Rare, medium or well-done? 三分熟、五分熟或者全熟的？
What would you like to go with your steak? 你要什么菜与你的牛排搭配？（问配菜时用）
Is that right? 这样对吗？

趣味阅读 Interesting Reading

牛排的生熟程度，在英文中一般可以分为五个等级：全熟是 well-done，七分熟是 mediumwell，五分熟是 medium，四分熟是 mediumrare，三分熟是 rare。

well-done就和头顶的感觉差不多，硬邦邦的；medium well 就是额头，稍微有点弹性；medium 是鼻头，软软的；rare就是下巴最柔软的部分。

比较常见的牛排有：菲力排、肉眼排、西冷排和T骨排。菲力排，取材于牛脊上最嫩的肉，几乎不含肥膘，肉嫩汁多；肉眼排，取材于牛背中部，瘦肉和肥肉兼而有之，由于含一定肥膘，这种肉煎烤味道比较香；西冷排，牛背后部的肉最为适合，有一定的

肥油，口感韧度强，不要煎得过熟；T骨排，呈T字形，是牛背上的脊骨肉。

鸡蛋的熟度和肉又不一样。对于要下油锅的鸡蛋，回答方法是①scramble炒蛋（就是全熟），②sunny side up只煎一面的荷包蛋（因为荷包蛋像太阳，所以美国人用sunny side 来形容），③sunny side down/easeover两面都煎。

     scramble   sunny side down /easeover   sunny side up

另外还有白煮蛋。这个分两种。汉语中是"嫩""老"之分，英语中是soft boil 和 hard boil。所谓的soft boil是指让蛋黄的部分还有点液体状，而hard boil则是指整个蛋黄都煮成固体状的。

    hard boil   soft boil

练习 Exercises

（1）以下三块牛排的烹饪程度分别为rare，medium，well-done。根据你自己的经验，判断牛排的熟度，并将对应的英文表达写在右边横线上。（How is the steak prepared?）

 _____

 _____

 _____

（2）如果你是店里的部门经理，你为店里的顾客们设计了这样一份问卷调查，以便以后烹制出更受顾客喜爱的菜品。请你寻找10位顾客，对他们进行问卷调查。请他们在适合自己的项目后面的括号处打√。根据课文，结合该问卷调查表，编一组小对话。（Questionnaire.）

Questionnaire

1. How would you like your steak prepared?

Rare　　　　　　　　　　　（　）

Medium　　　　　　　　　（　）

Well-done　　　　　　　　（　）

2. What soup do you prefer?

Mixed mushroom cream soup　（　）

Seafood consommé　　　　　（　）

3. What would you like to go with your steak?

Potato salad　　　　　　　（　）

Tomato salad　　　　　　　（　）

Fruit salad　　　　　　　　（　）

二、菜肴制作
The Cooking Ways

1. 汤 Soup

（W——Western Cook　C——Chinese Cook）

C: What are you doing?

W: I'm making soup.

C: What soup?

W: French onion soup.

C: The soup sounds rather fascinating. How do you prepare it?

W: First slice 40 onions. Then sauté the onions in butter.

C: Over low heat?

W: Yes, and stir occasionally.

C: For how long?

W: 20 or 30 minutes.

C: What next?

W: Add the beef broth, water, bay leaves, pepper and thyme. Then heat it to boiling. Reduce the heat. Cover and simmer. And in the meantime, you can toast the Baguette.

C: How long?

W: About 15 minutes. Then put the toasted Baguette slices into bowls…and poured onion soup on top.

C: Is that all?

W: No. The last step is to sprinkle Swiss and Parmesan cheese on top and put the bowls in the salamander just before serving.

C: Oh, I've got it. Thank you!

W: Not at all.

基础词汇 Basic Vocabulary

fascinating ['fæsineitiŋ] *adj.* 吸引人的

onion ['ʌnjən] *n.* 洋葱（头）

add [æd] *vt. & vi.* 加，加入；增加，添加

bay [bei] *n.* 月桂树

thyme [taim] *n.* （用以调味的）百里香（草）

cover ['kʌvə] *vi.* 覆盖；代替

meantime ['mi:ntaim] *n.* 其时，其间

sprinkle ['spriŋkl] *vt.* 洒；微雨；散置

salamander ['sælə,mændə] *n.* 烤箱

slice [slais] *vt.* 切

butter ['bʌtə] *n.* 黄油

broth [brɔθ] *n.* 肉汤

pepper ['pepə] *n.* 胡椒粉；辣椒

reduce [ri'dju:s] *vt.* 缩减，减少；降低

simmer ['simə] *n.* 炖 *vt. & vi.* 炖；慢煮

toast ['təust] *vi.* 烘，烤

cheese [tʃi:z] *n.* [食品] 奶酪；干酪

重点短语和句型 Key Phrases and Sentences

over low heat　慢火

reduce the heat　把火调小；降温

in the meantime   在此期间

Pour onion soup on top.   把洋葱汤浇到顶部。

pour... on top   把……浇/淋在顶部

What are you doing?   你在干吗？

How do you prepare it?   你是如何准备的？

What next?   下一步是什么？

How long?   多长？

Is that all?   这就是全部？这样就好了吗？

练习 Exercises

（1）为每幅图找出恰当的单词。（Find out the right word for each picture.）

①Chinese cook   ②Western cook   ③French onion soup

_____   _____   _____

（2）请你根据课文，帮助中国厨师写下这份食谱以及制作过程。（Write the recipe according to the dialogue for a Chinese cook.）

① 原材料：

② 制作过程：

2. 甜点 Dessert

(W——Western Cook   C——Chinese Cook)

W: The Black Forest Cake is very popular in western countries.

C: I've heard it. A western friend of mine once told me. Can you teach me how to make it?

W: Sure. First preheat the oven to 350°F(华氏摄氏度). And grease and flour two cake pans. Then put the flour, sugar, coca, baking soda, salt, baking powder, water, shortening, eggs and vanilla in the mixer.

C: Mix the ingredients at low speed?

W: Yes. For 30 seconds. Next you should mix the ingredients at high speed for 3 minutes. Then pour them into baking pans. And put the two pans in the oven for about 30 minutes. In the meantime you can prepare the cherry filling. After the cake is ready, put the cake on wire racks to cool. When it has cooled, you can spread the whipping cream, cherry filling and frosting, and garnish with chocolate curls and Maraschino cherries. It's done.

C: Oh, sounds great! I want to do it right away.

W: Sure. Go ahead.

基础词汇 Basic Vocabulary

preheat [ˌpriːˈhiːt] *vt.* 预热
flour [ˈflauə] *n.* 面粉 *vt.* 撒粉于；把…磨成粉
soda [ˈsəudə] *n.* 苏打，碱；苏打水，汽水
powder [ˈpaudə] *n.* 粉；粉末
shortening [ˈʃɔːtəniŋ] *n.* 起酥油
filling [ˈfiliŋ] *n.* 填充；填料
spread [spred] *vt. & vi.* 伸开，展开，摊开；(使)传播，(使)散布
whipping [ˈwipiŋ] cream 可打稠制作掼奶油的奶油
frosting [ˈfrɔstiŋ] *n.* 结霜；霜状白糖
garnish [ˈgɑːniʃ] *n.* 为增加色香味而添加的配菜
chocolate [ˈtʃɔkəlit] *n.* 巧克力；巧克力糖
maraschino [ˌmærəˈskiːnəu] *n.* 黑樱桃酒

grease [griːs] *vt.* 涂脂于
baking soda *n.* 碳酸氢钠，发酵粉
baking powder *n.* [食品] 发酵粉
ingredient [inˈgriːdiənt] *n.* 原料
wire [ˈwaiə] *n.* 金属丝，金属线
rack [ræk] *n.* 架子

curl [kəːl] *n.* 卷曲；卷发；螺旋状物
maraschino cherry 酒浸樱桃

重点短语和句型 Key Phrases and Sentences

Black Forest Cake   黑森林蛋糕
at low speed   低速   at high speed   高速

in the meantime　同时

right away　立刻，马上

go ahead　开始；干吧

I've heard it.　我曾经听说过这个。

练习 Exercises

根据对话写出黑森林蛋糕的制作方法。（The cooking way of the Black Forest Cake. ）

黑森林蛋糕

material：_____

_____

_____

_____

_____

Step1, _____

Step2, _____

Step3, _____

Step4, _____

Step5, after the cake is ready, _____

Step6, when it have cooled, _____

_____

It's done.

三、项目十一总练习
Exercises for part XI

Translate the following into English.

（1）我能先看一下菜单吗？
_____

（2）您想怎么做您的牛排？是三分熟、五分熟还是全熟？
_____

（3）加入牛肉高汤、月桂叶、水和百里香。
_____

（4）同时你可以烤法式面包。
_____

（5）将烤箱预热至350华氏度。
_____

（6）蛋糕烤好后放在架子上晾凉。
_____

Part IV Appendix
第四部分　附录

一、常见中式菜肴英文表述
The English Expressions of Common Chinese Dishes

冷菜 Cold Dishes

白灵菇扣鸭掌	Mushrooms with Duck Feet
白切鸡	Boiled Chicken with Sauce
冰梅凉瓜	Bitter Melon in Plum Sauce
冰镇芥蓝	Chinese Broccoli with Wasabi
朝鲜辣白菜	Korean Cabbage in Chili Sauce
夫妻肺片	Pork Lungs in Chili Sauce
酱香猪蹄	Pig Feet Seasoned with Soy Sauce
韭黄螺片	Sliced Sea Whelks with Hotbed Chives
老醋泡花生	Peanuts Pickled in Aged Vinegar
凉拌金针菇	Golden Mushrooms and Mixed Vegetables
卤水大肠	Marinated Pork Intestines
卤水金钱肚	Marinated Pork Tripe
萝卜干毛豆	Dried Radish with Green Soybean
蜜汁叉烧	Honey-Stewed BBQ Pork
明炉烧鸭	Roast Duck
蒜蓉海带丝	Sliced Kelp in Garlic Sauce
五香牛肉	Spicy Roast Beef
五香熏鱼	Spicy Smoked Fish
盐焗鸡	Baked Chicken in Salt
米醋海蜇	Jellyfish in Vinegar
桂花糯米藕	Steamed Lotus Root Stuffed with Sweet Sticky Rice
素鸭	Vegetarian Duck
酱鸭	Duck Seasoned with Soy Sauce
醉鸡	Liquor-Soaked Chicken
水晶鱼冻	Fish Aspic
小黄瓜蘸酱	Small Cucumber with Soybean Paste
琥珀核桃	Honeyed Walnuts
香椿鸭胗	Duck Gizzard with Chinese Toon
炸花生米	Fried Peanuts
德州扒鸡	Braised Chicken, Dezhou Style
蒜泥白肉	Pork with Garlic Sauce

热菜 Hot Dishes

◎ Part A

东坡方肉	Braised Dongpo Pork
咕噜肉	Gulaorou（Sweet and Sour Pork with Fat）
红烧狮子头	Stewed Pork Ball in Brown Sauce
脆皮乳猪	Crispy BBQ Suckling Pig
回锅肉片	Sautéed Sliced Pork with Pepper and Chili
椒盐肉排	Spare Ribs with Spicy Salt
京酱肉丝	Sautéed Shredded Pork in Sweet Bean Sauce
毛家红烧肉	Braised Pork, Mao's Family Style
蜜汁火方	Braised Ham in Honey Sauce
蜜汁烧小肉排	Stewed Spare Ribs in Honey Sauce
糖醋排骨	Sweet and Sour Spareribs
咸鱼蒸肉饼	Steamed Pork and Salted Fish Cutlet
蟹汤红焖狮子头	Steamed Pork Ball with Crab Soup
鱼香肉丝	Yu-Shiang Shredded Pork（Sautéed with Spicy Garlic Sauce）
糖醋里脊	Fried Sweet and Sour Tenderloin（Lean Meat）
罗汉肚	Pork Tripe Stuffed with Meat
水晶肘	Stewed Pork Hock
九转大肠	Braised Intestines in Brown Sauce
清炸里脊	Deep-Fried Pork Filet
软炸里脊	Soft-Fried Pork Filet
尖椒炒肥肠	Fried Pork Intestines with Hot Pepper
米粉排骨	Steamed Spare Ribs with Rice Flour
蚂蚁上树	Sautéed Vermicelli with Spicy Minced Pork
冬笋炒肉丝	Sautéed Shredded Pork with Bamboo Shoots
炸肉茄盒	Deep-Fried Eggplant with Pork Stuffing
清蒸猪脑	Steamed Pig Brains
芋头蒸排骨	Steamed Spare Ribs with Taro
板栗红烧肉	Braised Pork with Chestnuts
蛋黄狮子头	Stewed Meat Ball with Egg Yolk

◎ Part B

XO酱炒牛柳条	Sautéed Beef Filet in XO Sauce
彩椒牛柳	Sautéed Beef Filet with Bell Peppers
白灼肥牛	Scalded Beef

番茄炖牛腩	Braised Beef Brisket with Tomato
干煸牛肉丝	Sautéed Shredded Beef in Chili Sauce
杭椒牛柳	Sautéed Beef Filet with Hot Green Pepper
黑椒牛排	Pan-Fried Beef Steak with Black Pepper
红酒烩牛尾	Braised Oxtail in Red Wine
水煮牛肉	Poached Sliced Beef in Hot Chili Oil
铁板牛肉	Sizzling Beef Steak
豉汁牛仔骨	Steamed Beef Ribs in Black Bean Sauce
陈皮牛肉	Beef with Dried Orange Peel
红烧牛蹄筋	Braised Beef Tendon in Brown Sauce

◎ Part C

葱爆羊肉	Sautéed Lamb Slices with Scallion
红焖羊排	Braised Lamb Chops with Carrots
烤羊腿	Roast Lamb Leg
涮羊肉	Mongolian Hot Pot
红焖羊肉	Stewed Lamb in Brown Sauce

◎ Part D

板栗焖仔鸡	Braised Chicken with Chestnuts
脆皮鸡	Crispy Chicken
大煮干丝	Braised Shredded Chicken with Ham and Dried Tofu
芙蓉鸡片	Sautéed Chicken Slices in Egg-White
宫保鸡丁	Kung Pao Chicken
沙茶鸡煲	Chicken with BBQ Sauce en Casserole
铁板掌中宝	Sizzling Chicken Feet
鸡丁核桃仁	Stir-Fried Diced Chicken with Walnuts
怪味鸡丝	Special Flavored Shredded Chicken
口水鸡	Steamed Chicken with Chili Sauce
贵妃鸡	Chicken Wings and Legs with Brown Sauce
叫化鸡	Beggars Chicken（Baked Chicken）
纸包鸡	Deep-Fried Chicken in Tin Foil
全聚德烤鸭	Quanjude Roast Duck
虫草炖老鸭	Stewed Duck with Aweto
辽参扣鹅掌	Braised Goose Feet with Sea Cucumber
香煎鹅肝	Pan-Fried Goose Liver
天麻炖乳鸽	Stewed Pigeon with Gastrodia Tuber

咸鸭蛋	Salted Duck Egg
煮鸡蛋	Boiled Egg
荷包蛋	Poached Egg
韭菜炒鸡蛋	Scrambled Egg with Leek
蛤蜊蒸蛋	Steamed Egg with Clams
蛋羹	Egg Custard

◎ Part E

干锅茶树菇	Griddle Cooked Tea Tree Mushrooms
香菇扒菜胆	Braised Vegetable with Black Mushrooms
鲍汁花菇	Braised Mushrooms in Abalone Sauce
白灵菇扒鲍片	Braised Sliced Abalone with Mushrooms
龙井金元鲍	Braised Abalone with Longjing Tea
鲍汁葱烧辽参	Braised Sea Cucumber in Abalone Sauce
鲍汁扣白灵菇	Braised Mushrooms in Abalone Sauce
鲍汁牛肝菌	Braised Boletus in Abalone Sauce
北极贝刺身	Scallops Sashimi
夏果澳带	Sautéed Scallops with Macadamia
多宝鱼（清蒸，豉汁蒸过桥）	Turbot（Steamed / Steamed with Black Bean Sauce / Boiled）
鳜鱼（清蒸，油浸，松子炸）	Mandarin Fish（Steamed / Fried / Deep-Fried with Pine Nuts）
松鼠鳜鱼	Sweet and Sour Mandarin Fish
海鲜脆皮豆腐	Fried Tofu with Seafood
鲈鱼（清蒸，红烧，锅仔泡椒煮）	Perch（Steamed / Braised / Stewed with Marinated Chili）
炭烧鳗鱼	BBQ Eel
雪菜墨鱼丝	Sautéed Shredded Cuttlefish with Potherb Mustard
白灵菇扣裙边	Braised Mushrooms with Turtle Rim
膏蟹（花雕酒蒸，姜葱炒，椒盐）	Green Crab（Steamed in Yellow Wine / Fried with Ginger and Scallion / Fried with Spicy Salt）
红烧甲鱼	Braised Turtle in Brown Sauce
肉蟹（姜葱炒，清蒸，椒盐）	Hardshell Crab（Stir-Fried with Ginger and Scallion / Steamed / Fried with Spicy Salt）
香辣蟹	Sautéed Crab in Hot Spicy Sauce
杭椒虾爆鳝	Sautéed Shrimps and Eel with Hot Green Pepper
清炒鳝糊	Sautéed Mashed Eel

澳洲龙虾（刺身，上汤焗，椒盐）	Australian Lobster (Sashimi / in Chicken Consommé / Fried with Spicy Salt)
翡翠虾仁	Sautéed Shrimps with Broccoli
干烧大虾	Dry-Braised Prawn with Ham and Asparagus
清炒水晶河虾仁	Sautéed Shelled River Shrimps
泰式辣椒炒虾仁	Sautéed Shrimps with Chili, Thai Style
银杏百合炒虾球	Sautéed Shrimp Balls with Lily Bulbs and Ginkgo
煎银鳕鱼	Pan-Fried Codfish Filet
松菇银鳕鱼	Braised Codfish with Mushrooms
泰汁煎银鳕鱼	Pan-Fried Codfish in Thai Sauce
XO酱花枝片	Sautéed Sliced Squid in XO Sauce
铁板酱鲜鱿	Sizzling Fresh Squid with Soy Sauce
菜心扒鱼圆	Braised Fish Balls with Shanghai Greens
砂锅鱼头	Fish Head en Casserole
酸菜鱼	Boiled Fish with Pickled Cabbage and Chili
西湖醋鱼	West Lake Fish in Vinegar Gravy
浓汁鱼肚	Braised Fish Maw with Chicken Broth
虾子大乌参	Braised Jumbo Sea Cucumber with Shrimp Roe
清蒸石斑鱼	Steamed Sea Bass
虾仁跑蛋	Shrimp Omelet
油爆虾	Stir-Fried Shrimps
咖喱焗肉蟹	Baked Hardshell Crab with Curry
咸蛋黄珍宝蟹	Sautéed Jumbo Crab with Salted Egg Yolk
清蒸白鳝	Steamed Eel
白灼生中虾	Scalded Prawns
蒜蓉开边蒸生中虾	Steamed Prawns with Chopped Garlic
奶油焗龙虾	Baked Lobster with Fresh Cream
百合虾球	Sautéed Prawn with Lily Bulbs
水煮明虾	Poached Prawns
上汤焗龙虾	Poached Lobster in Chicken Consommé
清蒸武昌鱼	Steamed Wuchang Fish
红烧带鱼	Braised Ribbonfish in Brown Sauce
清蒸大闸蟹	Steamed Dazha Crabs
醉蟹	Liquor-Soaked Crabs

◎ Part F

百合炒南瓜	Sautéed Pumpkin with Lily Bulbs

炒芥蓝	Sautéed Chinese Broccoli
炒生菜	Sautéed Lettuce
炒时蔬	Sautéed Seasonal Vegetable
冬菇扒菜心	Braised Shanghai Greens with Black Mushrooms
蚝油生菜	Sautéed Lettuce in Oyster Sauce
红烧毛芋头	Braised Taro in Brown Sauce
虎皮尖椒	Pan-Seared Green Chili Pepper
萝卜干炒腊肉	Sautéed Preserved Pork with Dried Turnip
浓汤娃娃菜	Stewed Baby Cabbage in Broth
上汤浸时蔬	Braised Seasonal Vegetable in Broth
番茄炒蛋	Scrambled Egg with Tomato
银杏炒百合	Sautéed Lily Bulbs with Gingko
鱼香茄子煲	Yu-Shiang Eggplant in Casserole (Sautéed with Spicy Garlic Sauce)
鸭黄焗南瓜	Braised Pumpkin with Salted Egg Yolk
开水白菜	Steamed Chinese Cabbage in Supreme Soup
炝黄瓜	Spicy Cucumbers
蔬菜沙拉	Vegetables Salad
四川泡菜	Pickles, Sichuan Style
地三鲜	Sautéed Potato, Green Pepper and Eggplant

◎ Part G

脆皮豆腐	Deep-Fried Tofu
锅煽豆腐	Tofu Omelet
家常豆腐	Fried Tofu, Home Style
麻婆豆腐	Mapo Tofu (Stir-Fried Tofu in Hot Sauce)
百叶包肉	Tofu Skin Rolls with Minced Pork

汤羹类 Soups

冰糖银耳燕窝	Braised Bird's Nest with White Fungus and Rock Candy
一品燕窝	Best Quality Bird's Nest Soup
太极素菜羹	Thick Vegetable Soup
西湖牛肉豆腐羹	Minced Beef and Tofu Soup
番茄蛋花汤	Tomato and Egg Soup
龙虾浓汤	Lobster Soup
萝卜丝鲫鱼	Crucian Carp Soup with Shredded Turnips
雪菜大汤黄鱼	Stewed Yellow Croaker with Bamboo Shoots and Potherb Mustard

煲类 Casserole

锅仔药膳乌鸡	Stewed Black-Boned Chicken with Chinese Herbs
砂锅鱼头豆腐	Stewed Fish Head with Tofu in Pottery Pot
梅菜扣肉煲	Steamed Pork with Preserved Vegetable in Casserole

主食、小吃类 Rice, Noodles and Local Snacks

米饭	Steamed Rice
八宝饭	Eight Delicacies Rice
咖喱猪排饭	Rice with Spare Ribs in Curry Sauce
咖喱鸡饭	Rice with Curry Chicken
海鲜乌冬汤面	Japanese Soup Noodles with Seafood
榨菜肉丝汤面	Noodle Soup with Preserved Vegetable and Shredded Pork
拉面	Hand-Pulled Noodles
北京炸酱面	Noodles with Soy Bean Paste, Beijing Style
担担面	Noodles, Sichuan Style
海虾云吞面	Noodles with Fresh Shrimp Wonton
红烧牛腩面	Braised Beef Brisket Noodles
凉面	Cold Noodles with Sesame Sauce
阳春面	Plain Noodles
过桥肥牛汤米线	Vermicelli in Soup with Beef
干炒牛河	Stir-Fried Rice Noodles with Beef
蒸肠粉	Steamed Rice Rolls
酸辣粉	Hot and Sour Rice Noodles
猪肉白菜水饺	Jiaozi Stuffed with Pork and Chinese Cabbage
四喜鸭蓉饺	Jiaozi Stuffed with Duck Meat
蒸饺	Steamed Jiaozi
叉烧包	Steamed BBQ Pork Bun
生煎包	Pan-Fried Bun Stuffed with Pork
奶黄包	Steamed Creamy Custard Bun
三鲜小笼包	Steamed Bun Stuffed with Three Fresh Delicacies
小笼汤包	Steamed Bun Stuffed with Juicy Pork
葱油饼	Fried Baked Scallion Pancake
肉末烧饼	Sesame Cake with Minced Pork
烧卖	Shaomai (Steamed Pork Dumplings)
脆皮春卷	Crispy Spring Rolls

锅贴	Guotie（Pan-Fried Meat Dumplings）
绿豆粥	Congee with Mung Bean Congee
莲蓉酥	Lotus Seed Puff Pastry
蛋挞	Egg Tart
汤圆	Tangyuan（Glutinous Rice Balls）
五香茶叶蛋	Tea Flavored Boiled Eggs
羊肉泡馍	Pita Bread Soaked in Lamb Soup
豌豆黄	Pea Cake
炸麻球	Deep-Fried Glutinous Rice Balls with Sesame
烤银丝卷	Baked Steamed Rolls
麻薯	Deep-Fried Glutinous Rice Cake Stuffed with Bean Paste
寿桃	Birthday Bun with Bean Paste Filling
南瓜酪	Pumpkin Pastry
艾窝窝	Aiwowo（Steamed Rice Cakes with Sweet Stuffing）
驴打滚	Lúdagunr（Glutinous Rice Rolls with Sweet Bean Flour）
油炸臭豆腐	Deep-Fried Fermented Tofu
豆浆	Soybean Milk
豆腐脑儿	Jellied Tofu
酒酿圆子	Boiled Glutinous Rice Balls in Fermented Glutinous Rice
杏仁豆腐	Almond Jelly
拔丝山药	Chinese Yam in Hot Toffee

二、常见西式菜肴英文表述
The English Expressions of Common Western Dishes

头盘及沙拉类 Appetizers, Starters and Salads

腌熏三文鱼	Smoked Salmon
凯撒沙拉	Caesar Salad
金枪鱼沙拉	Tuna Salad

汤类 Soups

奶油蘑菇汤	Cream of Mushroom Soup

墨西哥辣味牛肉汤	Mexican Chili Beef Soup
番茄浓汤	Tomato Bisque Soup
法式洋葱汤	French Onion Soup
香浓牛尾汤	Oxtail Soup
意大利蔬菜汤	Minestrone Soup

禽蛋类 Poultry and Eggs

红酒鹅肝	Braised Goose Liver (Foie Gras in Red Wine)
炸培根鸡肉卷	Deep-Fried Chicken and Bacon Rolls
咖喱鸡	Chicken Curry
秘制鸭胸配黑菌炒马铃薯	Pan-Fried Duck Breast with Sautéed Potatoes and Truffles
火腿煎蛋	Fried Egg with Ham
洛林乳蛋饼	Quiche Lorraine
熘糊蛋	Scrambled Egg

牛肉类 Beef

红烩牛肉	Beef Stew
牛里脊扒配黑椒少司	Grilled Beef Tenderloin with Black Pepper Sauce
T骨牛扒	T-Bone Steak
铁板西冷牛扒	Sizzling Sirloin Steak
烩牛舌	Braised Ox Tongue

猪肉类 Pork

| 意大利米兰猪排 | Pork Piccatta |
| 炸猪排 | Fried Spire Ribs |

羊肉类 Lamb

| 扒新西兰羊排 | Grilled New Zealand Lamb Chops |
| 烤羊排配奶酪和红酒汁 | Roast Lamb Chops in Cheese and Red Wine Sauce |

鱼和海鲜类 Fish and Seafood

三文鱼扒配青柠黄油	Grilled Salmon with Lime and Butter
煎比目鱼	Pan-Fried Flatfish
煎红加吉鱼排	Grilled Sea Bream Filet
蒜蓉大虾	Grilled King Prawns with Garlic, Herb and Butter
巴黎黄油烤龙虾	Baked Lobster with Garlic and Butter

面、粉及配菜类 Noodles, Pasta and Side Dishes

海鲜通心粉	Macaroni with Seafood
海鲜意粉	Spaghetti with Seafood
什莱奶酪比萨饼	Vegetarian Pizza
肉酱意大利粉	Spaghetti Bolognaise
咖喱海鲜炒饭	Stir-Fried Seafood Rice with Curry
美式热狗	American Hot Dog
烤牛肉三文治	Roast Beef Sandwich
薯泥	Mashed Potato

面包类 Bread and Pastries

水果丹麦	Fruit Danish
牛角包	Croissant
全麦包	Whole Wheat Bread
长法棍	French Baguette
吐司面包	Toast

甜品及其他西点 Cakes, Cookies and Other Desserts

黑森林蛋糕	Black Forest Cake
草莓奶酪蛋糕	Strawberry Cheese Cake
意大利提拉米苏	Italian Tiramisu
干果派	Mixed Nuts Pie
芒果木司蛋糕	Mango Mousse Cake
杏仁多纳圈	Almond Donuts
水果蛋挞	Fresh Fruit Tart
绿茶布丁	Green Tea Pudding

香草布丁	Vanilla Pudding
蝴蝶酥	Butterfly Cookies
巧克力曲奇	Chocolate Cookies
爆米花（甜/咸）	Popcorn（Sweet / Salty）
香草冰淇淋	Vanilla Ice Cream
巧克力冰淇淋	Chocolate Ice Cream
冰霜	Sherbet

三、课文相关练习答案
The Answers to the Questions in the Textbook

一、菜名英文表述的基本方法概述
The Outline of the English Expressions of Chinese Dishes

学生学习 Student Learning

导入 Lead-in

（1）结合图片找出下列菜肴的主料、辅料、汤汁分别是什么，填入表格。

	主料	辅料	汤汁
玉米肉丸	肉丸	玉米	/
姜汁鲜鱿	新鲜鱿鱼	/	姜汁
美味牛筋	筋腱	/	/
麻辣肚丝	猪肚	/	麻辣汁
皮蛋豆腐	豆腐	皮蛋	/

（2）从给出的例子中可以看出，一般菜肴若存在辅料和汤汁，那主料、辅料、汤汁的顺

序一般为主料—辅料—汤汁，而他们两两之间均用介词（一般为in / and / with）隔开，因此可得出菜肴英文名的基本表达为：

主料（名称 / 形状）+ 介词 + 辅料 + 汤汁

基础公式学习 Basic Expressions Learning

◎ 巩固练习 Exercise to Consolidate

in / with / and

练习 Exercises

桂花山药	Chinese Yam with Osmanthus Sauce
豆豉鲫鱼	Crucian Carp with Black Bean Sauce
酸辣瓜条	Cucumber with Hot and Sour Sauce
香辣茄子	Eggplant with Chili Oil
姜汁蜇皮	Jellyfish in Ginger Sauce

二、常用烹饪方法的英文表述
The English Expressions of Common Chinese Cookery

学生学习 Student Learning

导入 Lead-in

常用烹饪方法的英文表达 The English Expressions of Common Chinese Cookery

◎ 巩固练习 Exercise to Consolidate

① 连一连（Match Task）

请帮忙将逃跑的气球固定在正确的位置上。

炒：sauté　煮：boil　酱：marinate　酿：ferment　腌：pickle　蒸：steam

② 以下单词中，哪几个单词与下图厨房用具有关，请将相关的单词放入锅中方框处。

simmer / stew / braise

公式学习 Expressions Learning

◎ 思考题 Questions

① 规律：用的都是动词过去分词形式；位置都在菜名的首位。

② 菜名的构成除了烹饪技法外，仍然符合基本公式的规律。

◎ 练习 Exercises

结合以上图片，分组讨论以下菜肴主要运用了什么烹饪技法，用所学英文单词表示。

卤水大肠	Marinate	浓汤娃娃菜	Stew

烟熏蜜汁肋排	Smoke	清蒸火腿鸡片	Steam
酸黄瓜条	Pickle	烤银丝卷	Bake
干豆角回锅肉	Sauté	烤全羊	Roast
萝卜干腊肉	Sauté	白灼西蓝花	Scald
当红炸子鸡	Deep-Fry	香辣虾	Fry
孜然烤牛肉	Grill	火爆川椒鸭舌	Stir-Fry
鲍鱼红烧肉	Braise	炸肉茄盒	Deep-Fry
药味炖生中虾	Simmer	烫青菜	Scald
冬菜扣肉	Braise	鲜奶煮蛋	Boil

拓展学习 Extended Learning

练习 Exercises

椰汁鲜果西米露	<u>Sweet</u> Sago with Fresh Fruit in Coconut Milk
五香牛肉	<u>Spicy</u> Roast Beef
薯泥	<u>Mashed</u> Potato
肉松松饼	<u>Shredded</u> Pork Puff
糖醋咕噜虾球	<u>Sweet and Sour</u> Prawn
香酱爆鸭丝	<u>Sautéed Shredded</u> Duck in Soy Sauce
回锅肉片	Sautéed <u>Sliced</u> Pork with Pepper and Chili
毛氏红烧肉	Chairman <u>Mao's</u> Stewed Pork with Soy Sauce

三、主要烹饪原料的英文表述
The English Expressions of Cooking Material

学生学习 Student Learning

烹饪原料——蔬菜类 Cooking Material-Vegetables

（1）豆类（Beans and Peas）

（2）瓜类；甜瓜（Melon）

◎ 巩固练习 Exercise to Consolidate

国际部餐厅负责进货的是个英国小伙，你希望他帮助你进以下原料，请告诉他你需要原料的英文名称，并给出最新的市场价以防他买贵。

cucumber / white gourd / pumpkin / bitter melon / papaya

（3）绿叶蔬菜（Green Vegetables）

◎ 巩固练习 Exercise to Consolidate

spinach / celery / Chinese broccoli / green vegetable / cabbage / lettuce / leek / broccoli / asparagus / coriander

（4）菌菇类（Mushrooms）

◎ 巩固练习 Exercise to Consolidate

tea tree mushrooms / golden mushrooms / mushroom / boletus / straw mushrooms

◎ 巩固练习 Exercise to Consolidate

根据前面所学，翻译菜名。

百合炒南瓜	Sautéed Pumpkin with Lily Bulbs
炒芥蓝	Sautéed Chinese Broccoli
翠豆玉米粒	Sautéed Green Peas and Sweet Corn
冬菇扒菜心	Braised Shanghai Greens with Black Mushrooms
火腿炒蚕豆	Sautéed Broad Beans with Ham
木瓜炖百合	Stewed Papaya with Lily Bulbs
清煎番茄	Pan-Fried Tomato
番茄炒蛋	Scrambled Egg with Tomato
烫青菜	Scalded Green Vegetable
青叶豆腐	Steamed Tofu with Green Vegetable
白灼西蓝花	Scalded Broccoli
脆皮豆腐	Deep-Fried Tofu
牛肝菌红烧豆腐	Braised Tofu with Boletus
雪菜炒豆皮	Sautéed Tofu Skin with Potherb Mustard
子姜鸭	Sautéed Duck with Ginger Shoots
清炒 / 蒜蓉西蓝花	Sautéed Broccoli with / without Garlic

烹饪原料——肉类 Cooking Material-Meat

烹饪中常用肉类（Meat in Common Use）

◎ 巩固练习 Exercise to Consolidate

一个粗心的服务生，把下面家禽的肉的英文名称标签贴错了，你能将这些打乱的标签，重新贴到对应的动物上吗？（填写到空格处）

goose / chicken / beef / duck / mutton / pork

◎ 巩固练习 Exercise to Consolidate

根据前面所学，翻译菜名。

冬菜扣肉	Braised Pork with Preserved Vegetables

腐乳猪蹄	Stewed Pig Feet with Preserved Tofu
爆炒牛肋骨	Sautéed Beef Ribs
番茄炖牛腩	Braised Beef Brisket with Tomato
芫爆散丹	Sautéed Ox Tripe with Coriander
烤羊里脊	Roast Lamb Tenderloin
烤羊腿	Roast Lamb Leg
可乐凤中翼	Pan-Fried Chicken Wings in Coca-Cola Sauce
鸿运蒸凤爪	Steamed Chicken Feet
火燎鸭心	Sautéed Duck Hearts
黑椒焖鸭胗	Braised Duck Gizzards with Black Pepper
葱爆肥牛	Sautéed Beef with Scallion

烹饪原料——水产品类 Cooking Material-Fishery Products

◎ 巩固练习 Exercise to Consolidate

根据前面所学，翻译菜名。

鲍鱼烧牛头	Braised Abalone with Ox Head
鲍汁豆腐	Braised Tofu in Abalone Sauce
干煎带鱼	Deep-Fried Ribbonfish
清蒸鳜鱼	Steamed Mandarin Fish
豆腐烧鱼	Fried Fish with Tofu
茄汁虾仁	Sautéed Shrimps with Tomato Sauce
生炒鳗片	Sautéed Eel Slices
鲶鱼烧茄子	Braised Cat Fish with Eggplant
鲜豌豆炒河虾仁	Sautéed Shrimps with Fresh Beans
冬菜银鳕鱼	Steamed Codfish with Preserved Vegetables
清蒸白鳝	Steamed Eel
白灼生中虾	Scalded Prawns
子姜虾	Sautéed Shrimps with Ginger Shoots
大蒜烧白鳝	Braised Eel with Garlic

烹饪原料——蛋制品 Cooking Material-Egg Products

◎ 巩固练习 Exercise to Consolidate

蛋黄明虾	Deep-Fried Prawns with Egg Yolk
木瓜牛奶蛋黄汁	Papaya Milk with Egg Yolk
皮蛋豆腐	Tofu with Preserved Eggs

| 口蘑煎蛋卷 | Mushroom Omelets |

烹饪原料——主食和小吃 Cooking Material-Staple Food and Snacks

◎ 巩固练习 Exercise to Consolidate

① 写出下列主食和小吃的英文名。

pancake / egg tart / Guotie / steamed twisted roll / rice flour / puff pastry / Youtiao / Wonton / puff / sticky rice; glutinous rice / Wotou

② 根据前面所学，翻译菜名。

米饭	Steamed Rice
海皇炒饭	Fried Rice with Seafood
鸡汤面	Chicken Noodle Soup
红烧牛腩汤面	Soup Noodles with Beef Brisket
红烧排骨汤面	Soup Noodles with Spare Ribs
煎包	Pan-Fried Dumplings
香滑芋蓉包	Steamed Taro Bun
韭菜晶饼	Steamed Leek Pancake
萝卜丝酥饼	Pan-Fried Turnip Cake
芋丝炸春卷	Deep-Fried Taro Spring Rolls
黑米小窝头	Wotou with Black Rice (Steamed Black Rice Bun)
小米金瓜粥	Millet Congee with Pumpkin
绿豆粥	Congee with Mung Bean Congee
炸云吞	Deep-Fried Wonton

烹饪原料——水果类 Cooking Material-Fruits

◎ 巩固练习 Exercise to Consolidate

① 用英文为下列水果命名。

lemon / litchi / sugarcane / strawberry / watermelon

② 将下列水果翻译成中文并试着画出这些水果。

龙眼、菠萝、火龙果、猕猴桃、荸荠

③ 根据前面所学，翻译菜名。

明虾荔枝沙拉	Shrimps and Litchi Salad
荔枝炒牛肉	Sautéed Beef with Litchi
淮山圆肉炖甲鱼	Braised Turtle with Yam and Longan
菠萝虾球	Sautéed Prawn with Pineapple
冰梅凉瓜	Bitter Melon in Plum Sauce

雪豆马蹄	Snow Peas with Water Chestnuts
松仁香菇	Black Mushrooms with Pine Nuts
板栗红烧肉	Braised Pork with Chestnuts
花生糕	Peanut Cake
鸡丁核桃仁	Stir-Fried Diced Chicken with Walnuts
雪梨炖百合	Snow Pear and Lily Bulbs Soup
杏仁炒南瓜	Sautéed Pumpkin with Almonds

烹饪原料——干货制品 Cooking Material-Dried Products

◎ 巩固练习 Exercise to Consolidate

鲍汁扣辽参	Braised Sea Cucumber in Abalone Sauce
蛋花炒鱼肚	Sautéed Fish Maw with Scrambled Egg
芝麻炸多春鱼	Deep-Fried Shisamo with Sesame
木瓜腰豆煮海参	Braised Sea Cucumber with Kidney Beans and Papaya
枸杞蒸裙边	Steamed Turtle Rim in Chinese Wolfberry Soup
牛肝菌红烧豆腐	Braised Tofu with Boletus
鲜人参炖土鸡	Braised Chicken with Ginseng
燕窝鸽蛋	Bird's Nest with Pigeon Egg
冰糖银耳炖雪梨	Stewed Pear with White Fungus and Rock Candy
萝卜干毛豆	Dried Radish with Green Soybean
白汁烧裙边	Stewed Turtle Rim
鲍汁扣花胶皇	Braised Fish Maw in Abalone Sauce

四、常用烹饪调料的英文表述
The English Expressions of Common Condiment

学生学习 Student Learning

基础词汇学习 Basic Words Learning

◎ 巩固练习 Exercise to Consolidate

sugar / salt / vinegar / pepper / honey

常用调味汁和少司 The Common Sauce

◎ 巩固练习 Exercise to Consolidate

根据前面所学，翻译菜名。

孜然寸骨	Sautéed Spare Ribs with Cumin

香草蒜蓉炒鲜蘑	Sautéed Fresh Mushrooms with Garlic and Vanilla
冰糖甲鱼	Steamed Turtle in Crystal Sugar Soup
冰花炖官燕	Braised Bird's Nest with Rock Candy
香酱爆鸭丝	Sautéed Shredded Duck in Soy Sauce
姜汁皮蛋	Preserved Eggs in Ginger Sauce
蒜蓉海带丝	Sliced Kelp in Garlic Sauce
咖喱猪排饭	Rice with Spare Ribs in Curry Sauce
冰梅凉瓜	Bitter Melon in Plum Sauce
蚝汁辽参扣鸭掌	Braised Sea Cucumber with Duck Feet in Oyster Sauce
葱油泼多宝鱼	Steamed Turbot with Scallion Oil
麻酱笋条	Shredded Lettuce with Sesame Paste

酒香类 Wine

◎ 巩固练习 Exercise to Consolidate

① 在一场婚礼上，一个外国客人看到餐桌上的各种酒非常感兴趣。请向他介绍这些酒。

red wine / white wine / rice wine / Chinese Liquor / beer / yellow wine / aged rice wine

② 根据前面所学，翻译菜名。

青椒牛柳	Sautéed Beef Filet with Hot Green Pepper
辣白菜炒牛肉	Sautéed Beef with Cabbage in Chili Sauce
酸辣瓜条	Cucumber with Hot and Sour Sauce
芥末木耳	Black Fungus with Mustard Sauce
京酱龙虾球	Lobster Balls in Sweet Bean Sauce
豆豉豆腐	Sautéed Tofu with Black Bean Sauce
红酒鹅肝	Braised Goose Liver / Foie Gras in Red Wine
糟香鹅掌	Braised Goose Feet in Rice Wine Sauce
啤酒鸡	Stewed Chicken in Beer

五、常用菜肴风味的英文表述
The English Expressions of Common Flavors

学生学习 Student Learning

公式学习 Expressions Learning

◎ 练习 Exercises

根据前面所学和以上图片，翻译菜名。

川汁牛柳	Sautéed Beef Filet in Chili Sauce, Sichuan Style
湖南牛肉	Beef, Hunan Style
潮州烧雁鹅	Roast Goose, Chaozhou Style
上海油爆虾	Stire-Fried Shrimps, Shanghai Style
上海菜煨面	Noodles with Vegetables, Shanghai Style
日式蒸豆腐	Steamed Tofu, Japanese Style
中式牛排	Beef Steak with Tomato Sauce, Chinese Style
台式蛋黄肉	Steamed Pork with Salted Egg Yolk, Taiwan Style
家常豆腐	Fried Tofu, Home Style

六、菜名的英文表述小结
The Summary of the English Expressions of Chinese Dishes

学生学习 Student Learning

公式学习 Expressions Learning

◎ 练习 Exercises

砂锅白菜粉丝	Chinese Cabbage and Vermicelli in Pottery Pot
石钵蟹黄豆腐	Stewed Crab Roe and Tofu in Stone Pot
砂锅鱼头豆腐	Stewed Fish Head with Tofu in Pottery Pot
铁锅牛柳	Braised Beef Filet in Iron Pot
瓦罐山珍	Mushrooms in Pottery Pot

◎ 练习 Exercises

川北凉粉	Clear Noodles in Chili Sauce
朝鲜辣白菜	Korean Cabbage in Chili Sauce
陈皮兔肉	Rabbit with Tangerine Flavor
夫妻肺片	Pork Lungs in Chili Sauce
干拌牛舌	Ox Tongue in Chili Sauce
红心鸭卷	Sliced Duck Rolls with Egg Yolk
姜汁皮蛋	Preserved Eggs in Ginger Sauce
韭菜炒鸡蛋	Scrambled Egg with Leek
干拌顺风	Pig Ear in Chili Sauce

项目二　厨房常用设备与工具
Kitchen Utensil and Cooking Equipment

一、刀具的英文表述
The English Expressions of Cutting Tools

学生学习　Student Learning

导入　Lead-in

◎ 思考题　Questions

① 常见的刀具可参见知识背景介绍。与刀具有关的单词：knife，sharp，blunt，dangerous 等。

② 可参见知识背景介绍。

不同种类的刀具　Words-Learning: Different Kinds of Knives

（2）观察下列单词的特点，结合中文意思，总结出这一组刀具表达法的共性。

　　boning knife [去骨刀]　　　　carving knife [切肉刀]

　　paring knife [去皮刀]　　　　coring knife [去核刀]

　　slicing knife [片刀]　　　　　chopping knife [文武刀]

共性：这一组刀具的表达方式，都是由"动名词+knife"构成。其中，动名词由相对应的动词转变而来。

◎ 巩固练习　Exercise to Consolidate

请根据图片写出相应的刀具。

boning knife / carving knife；slicing knife / paring knife / coring knife / slicing knife

（3）观察下列单词的特点，结合中文意思，总结出这一组刀具表达法的共性。

　　meat knife [肉刀]　　　　　　fish knife [鱼刀]

　　fruit knife [水果刀]　　　　　vegetable knife [蔬菜刀]

共性：这一组刀具的表达方式，都是由"名词+knife"构成。要注意，这里的名词表示原材料，都是不可数名词。

（4）观察下面这组单词，它们又有什么特点呢？它们与上一组有关系吗？

　　bread knife [面包刀]　　　　cheese knife [奶酪刀]

　　tomato knife [番茄刀]　　　　sandwich knife [三明治刀]

共性：这一组刀具的表达方式，也是由"名词+knife"构成。与上一组相比，这一组刀具是针对单一类食材中具体某种食材的刀具，而上一组则是针对某一类食材的刀具。

> 试一试 Have a try
>
> egg knife / butter knife / potato knife

刀的不同部位 Different Parts of Knives

◎ 任务 Tasks

（1）（从上到下，从左到右）答案分别是：knife point；knife blade；knife edge；knife back；hilt。

（2）此题为开放题。刀的各个部分都是按照其功能或者是外形特征命名的。与这些命名相关的单词有很多，学生言之有理即可。

例如：sharp，clean，firm，heavy 等。

练习 Exercises

（1）chop [剁] / chip [削] / cut [切]

（2）根据所学知识，给图片中的食材选择合适的刀具，填在图片下方的横线上。

① vegetable knife；tomato knife

② meat knife；carving knife；boning knife

③ vegetable knife；paring knife；slicing knife

④ bread knife

（3）练习（2）图中的四种食材，分别适合哪种加工方式？请为它们选择合适的动词。

① cut ② chop；cut ③ chip ④ cut；chip

（4）猜猜下列图片是哪种刀。

Cleaver　Bread　Boning　Chefs　Carving　Paring

二、加热设备的英文表述
The English Expressions of Cookware

学生学习 Student Learning

导入 Lead-in

◎ 思考题 Questions

① pan

② 可参见知识背景介绍。

语言知识 Language Knowledge

◎ 巩固练习 Exercise to Consolidate

试着用刚学习过的单词命名下面的加热设备。

boiler / pan / oven / boiler /

pot / boiler / oven / pan /

stove / boiler / pot / oven

各类烹饪加热设备 Different Kinds of Cookware

◎ 巩固练习 Exercise to Consolidate

试着用刚学习过的单词再一次命名下面的加热设备。

stew boiler / frying pan / oven / egg boiler /

coffee pot / steam boiler / microwave oven / milk pan /

electric stove / steam boiler / crock pot / oven

练习 Exercises

（1）答案不唯一

例：①-④的加热设备分别可选用：egg boiler；pressure cooker；gas range；automatic rice cooker。

（2）参见课文内容。

三、盛器的英文表述
The English Expressions of Containing Equipment

学生学习 Student Learning

导入 Lead-in

◎ 思考题 Questions

① 可参见知识背景介绍。

② 图中的盛器分别是：plate，glass，bowl

形容词汇：clean, round, big, beautiful 等。答案不唯一。

各类用餐盛器 Different Kinds of Dishware

◎ 思考题1 Question1

结合你的生活经验想想看，这三种盛器有什么主要区别？

三种盛器的主要区别：plate为盘子，碟子，浅而平；dish则是较深的盘子；bowl是碗，相对于前两者而言口径较小且有一定深度。

◎ 思考题2 Question2

想想看，这里的soup pot [汤罐] 和上面的soup bowl [汤碗]两者在意思上有什么联系，又有什么不同？你能结合自己的实践经验举出另外的例子吗？

两者的联系：都是用于盛汤的盛器，即盛放的食材一样。不同点：盛器有区别。也就是说，同样的食材可以根据需求放在不同的盛器里。另例：cake plate 和cake dish。

各类饮品盛器 Different Kinds of Drink-ware

◎ 巩固练习 Exercise to Consolidate

根据图片中客人所点的饮料选择正确的饮品盛器并写出其英文名。

①-b）goblet　　②-c）tankard　　③-a）coffee cup

练习 Exercises

答案不唯一，符合烹饪常识即可。鼓励学生思维多样化。具体词汇可参见本节及前两节的语言知识。

四、辅助设备的英文表述
The English Expressions of Auxiliary Equipment

学生学习 Student Learning

导入 Lead-in

◎ 思考题 Questions

① 可参见知识背景介绍。

② 答案不唯一。如学生觉得不齐全，引导其给出合理的建议，如发明某种新设备等。

基础知识 Basic Knowledge

（1）洗涤设备（Washing Equipment）

◎ 巩固练习 Exercise to Consolidate

Match Task

洗瓶用品可以为kitchen sink, bottle brush, detergent

洗碗用品可以为kitchen sink, dish cloth, detergent, mop

（2）备餐设备（Food Preparing Equipment）

◎ 巩固练习 Exercise to Consolidate

Match Task

过年时，你们家买了一箱的虾，但是实在太多了，一家人吃不掉。可以放进下列哪些设备中？请将线连上。

refrigerated cabinet / freezer / freezing compartment

（4）食材加工设备（Ingredients Preparing Equipment）

◎ 巩固练习 Exercise to Consolidate

Match Task

在箭头上写出所需设备的英文名，以达到从左图到右图的效果。

依次是electric juicer，blender，noodle press，food slicer，potato masher，meat grinder

练习 Exercises

（1）四幅图分别是：cooker hood（排油烟机），automatic dryer（自动干燥机），ice machine（制冰机）和food slicer（切片机）。

（2）答案不唯一。可参见知识背景介绍并结合教师上课的补充内容进行讨论。

项目三 厨师岗位英语
English for the Cook on the Post

一、厨房岗位的英文表述
The English Expressions of the Post in the Kitchen

学生学习 Student Learning

练习 Exercises

根据所学内容，看看以下原材料或者物品由厨房哪个岗位的厨师负责？

Pastry Chef; Larder Chef; Vegetable Chef / Cook; Soup Cook; Dishwasher; Fish Cook; Breakfast Cook; Sauce Cook

二、厨师职责的英文表述
The English Expressions of the Responsibilities of a Cook

学生学习 Student Learning

练习 Exercises

1. 给下面几个职责归类。

2. 根据所学内容填空。

① Keep <u>hair</u> clean and covered.

② Keep fingernails short and well <u>maintained</u>.

③ Turn off the flame and gas <u>as soon as</u> dish is cooked. Keep the kitchen ventilated.

④ Use clean, running water to <u>rinse</u>.

⑤ Cooked above, <u>raw</u> below or separated refrigerators.

三、厨房安全与卫生的英文表述
The English Expressions of Kitchen Safety and Hygiene

学生学习 Student Learning

练习 Exercises

① Make the duties of all the <u>positions</u> clear.

② All the rules must be <u>carried out</u> strictly.

③ Clean up the unnecessary things, divide goods into groups and <u>label</u> them.

④ Make sure all staffs learn the rules in the kitchen <u>by heart</u> in various forms of trainings.

一、散餐；按菜单点菜
A La Carte

练习 Exercises

（1）If you are attending a job interview, here are some questions appearing in the test paper.

Let's see whether you can pass the test or not.

① 第二幅图片下面打钩。

② waiter / guests

③ Mapo Tofu / Kung Pao Chicken / Yu-Shiang Shredded Pork

（2）Put the dialogue in order.

b, d, a, c

（3）Make out a dialogue.

参照课文对话，将新的菜名带入。

二、自助餐
Buffet

练习 Exercises

（1）Which type of dining should Mr. Bean choose?

选择buffet。左图为buffet，右图为a la carte。

（2）Put the dialogue in order.

b, a, c

（3）Design a poster and make out a dialogue.

海报更多激发学生的想象力和创造力；对话请结合课文内容编写。

三、酒水
Beverage

练习 Exercises

（1）Where to put the ice?

冰块放入右边酒杯中。

（2）According to the dialogue, which one does the guest like better, straight up or on the rocks? Why? What about you?

The guest likes straight up better. Ice will spoil the taste.

学生的回答按实际情况定。

（3）Make out a dialogue.

参照课文。

四、项目四总练习
Exercises for Project IV

Translate the following sentences into English.

(1) I'd like to have a table by the window.

(2) I prefer dry wine.

(3) We'll give a 50% discount for the olds (age＞60) and kids under 1.4 meter high.

(4) I'm sorry, it's already reserved. / I'm sorry, this table is already reserved.

(5) How much for one person?

项目五　菜肴特色介绍
Introducing Specialties

一、川菜
Sichuan Cuisine

练习 Exercises

Suppose you are a boss of a Sichuan cuisine restaurant. Can you answer the questions below?

(1) The restaurant caters to both Chinese and western taste.

(2) The guests prefer Chinese food.

(3) We want to try some spicy and hot dishes.

(4) Sichuan cuisine is famous for its hot, spicy food.

(5) He recommend Kung Pao Chicken, Yu-Shiang Shredded Pork, Mapo Tofu and the crucian carp with chili bean sauce.

(6) No. Sichuan food is not always hot. Apart from hot food, we have sweet and sour food that many foreign friends like, light taste food and multi-flavoured food.

二、鲁菜
Shandong Cuisine

练习 Exercises

(1) Name these two dishes in English.

Yellow River Carp in Sweet and Sour Sauce / Braised Sea Cucumber with Scallion

(2) Put the dialogue in order.

1, 3, 5, 6, 7, 2, 4

三、淮扬菜
Huaiyang Cuisine

练习 Exercises

(1) Tell something about Yangzhou food according to the dialogue.

Yangzhou food is light; Suxi food is sweet; Xuhai food is salty and fresh; Jinling food is crisp and tender

(2) Find out the wrong dish.

Yellow River carp in sweet and sour sauce

四、项目五总练习
Exercises for Project V

Translate the following sentences into English.

(1) We cater to both Chinese and western taste.

(2) Which would you prefer, Chinese or western?

(3) We'll have Chinese food for a change today.

(4) Could you recommend us something special?

(5) Apart from hot food, we have sweet and sour food that many foreign friends like, light taste food and multi-flavoured food.

(6) Our restaurant mainly serves Cantonese cuisine, including Chaozhou, Shantou and Hakka dishes.

(7) It uses a large variety of distinct ingredients including the meat of snakes and cats.

(8) Huaiyang cuisine occupies an important position in Chinese cuisine.

(9) Yangzhou food is light while Suxi food is sweet. Being crisp and tender is the characteristic of Jinling food, whereas the flavour of Xuhai food is salty and fresh.

项目六 菜肴制作过程介绍
Introducing the Cooking Ways

一、中餐
Chinese Food

热菜 Hot Dishes

练习 Exercises

(1) Questions about Beijing Roast Duck.

① Beijing Roast Duck

② Tender and crisp

③ split open, dressed, scalded, dried, roasting

④ About 50 minutes

(2) Food diary.

参照课文重点字词句。

凉菜 Cold Dishes

练习 Exercises

(1) What is their occupation?

chef / waiter / waitress

(2) Name the hot dish and cold dish below.

Asparagus Salad / Yellow River Carp in Sweet and Sour Sauce

(3) How to make the asparagus salad?

First, cut 1 pound asparagus diagonally, bring 4 cups water to boil in saucepan. Then drop in asparagus, boil 1 minute, drain, rinse with cold water. Mix next four ingredients, the last step is to pour the dressing over the asparagus.

二、项目六总练习
Exercises for Project Ⅵ

Translate the following sentences into English.

(1) I've never tasted anything like it before.

(2) It's a little complicated to make a Beijing Roasted Duck.

(3) First, duck must be split open, dressed, scalded and dried.

(4) When roasting, it's better to use fruit tree branches as firewood to lend more flavor to the duck.

(5) I'm sorry I don't understand. I'll ask the chef to help you.

(6) First, cut 1 pound asparagus diagonally, bring 4 cups water to boil in saucepan.

(7) Then drop in asparagus, boil 1 minute, drain, rinse with cold water.

项目七　厨房介绍
Introducing the Kitchen

一、中式厨房
Chinese Kitchen

练习　Exercises

Answer the questions below according to the pictures.

（1）c　chopping block

（2）d　walk-in refrigerator

（3）b　casserole

（4）a　bamboo steamer

二、厨房设备与工具
Kitchen Utensil and Cooking Equipment

盛器　Dishware

练习　Exercises

（1）Fill in the blanks.

① greasy, terrible

② hard

③ First of all

（2）Translate these following sentences.

① Can you give me a hand with something in the kitchen?

② What do you want me to do?

③ Come to help me with the soup pots and the wine glasses.

④ Let me deal with them.

加热用具　Cookware

练习　Exercises

（1）Match Task.

①-a，②-b，③-e，④-d，⑤-c

（2）Answer the questions according to the dialogue.

① She needs <u>the cookware immediately.</u>

② Because she has <u>invited her parents to have dinner tomorrow.</u>

③ Not really. Actually she has no idea about it.

④ He suggests Mary to <u>buy a pressure cooker, a gas range, an automatic cooker, an oven and a boiler.</u>

⑤ It's good for boiling food.

辅助设备 Auxiliary Equipment

练习 Exercises

(1) Complete the dialogue according to the text.

refrigerator / eagle-eyed / automatic dryer new / saves / any more / what / ice breaker / wonderful / especially / electric juice / great / buying

(2) Match Task.

①-c　②-d　③-a　④-b

厨房设备 Kitchen Utensil

练习 Exercises

(1) Answer questions.

① Freezing compartment.

② Ice machine.

③ Ice breaker.

④ Oil smoker exhausting equipment.

(2) Introduce the kitchen according to the requirement.

结合新学的前三段对话。

项目七总练习
Exercises for Project Ⅶ

Translate the following sentences into English.

(1) Can you give me a hand with something in the kitchen?

(2) It's really hard to clean them.

(3) I need to buy lots of things to furnish my new kitchen.

(4) I think I need the cookware immediately because I have invited my parents to have dinner tomorrow.

(5) According to you situation, I suggest you to buy a pressure cooker, a gas range, and an automatic cooker.

(6) Many thanks! I'll follow you advice.

(7) How about drinking a glass of juice?

(8) Now please allow me to give some introduction.

(9) Oh, it's very spacious and tidy.

(10) This is the most advanced oil smoker exhausting equipment.

项目八 面试英语(烹饪)
The English Expressions Used for a Cook Job Interview

一、自我介绍
Self-introduction

练习 Exercises

Write an English resume about yourself.

参考课文相关句式和短语。

二、面试对话
Interview

练习 Exercises

(1) Answer the interview questions below.

① I choose cuisine as my major because I love cooking. I really enjoy creating a meal from various ingredients and watching my family or friends enjoy it. It gives me a real sense of satisfaction.

② I think cooking needs patience, carefulness and practice. Impatience can not make a good cook.

③ I want to be a successful chef.

④ Firstly, I will study as hard as I can. Secondly, I will try to practice more and communicate with others. Thirdly, I will pay attention to my comprehensive quality and improve my abilities.

(答案不唯一)

(2) Role play.

教师自评；句子参照课文。

三、项目八总练习
Exercises for Project VIII

Translate the following sentences into English.

(1) Would you please introduce yourself briefly?

(2) I'm an active boy and I like sports. Cooking and basketball are my favorites.

(3) What's your strongpoint to be as a cuisine major?

（4）Now I have realized my weakness and I am trying to improve it.

（5）I think you can overcome it.

（6）Thank you for your self-introduction.

（7）I'm a hard-working student and like to learn from others.

（8）Would you explain why you choose cuisine as your major?

（9）It gives me a real sense of satisfaction.

（10）What is your career purpose?

（11）I will study as hard as I can.

（12）I want to be a successful chef.

项目九　厨师岗位
The Cook on the Post

一、从学徒工开始做起
Start from a commis

练习 Exercises

（1）看下面的图片，试着找出哪些工作可以属于学徒工。再根据所学内容，试着写下这些工作是由什么职位负责的。(Look at all the pictures below and try to find out what kinds of jobs can belong to a commis. Then, try to write down the position which is in charge of each job according to what you have learnt.)

Job List（工作清单）

①③④

Position：① Larder Chef ② Larder Chef ③ Caller ④ Dishwasher（答案不唯一）

（2）试着思考更多学徒工可以做的工作，然后用英语说给搭档听。(Try to think about more jobs a commis can do and talk to your partner in English.)（答案略）

二、协作在厨房
Cooperate with others in the kitchen

练习 Exercises

试着根据我们刚学的对话，跟搭档编写对话。情况如下。(Try to make out a dialogue with

your partner, according to the dialogue we've just learnt. The situation is like this.)

答案不唯一，鼓励基础弱的同学完全根据对话会仿造就好，基础好的同学可以发散思维，自由发挥一些。

三、厨房准则
Dos and don'ts in the kitchen

练习 Exercises

（1）根据对话内容，帮助杰克列出他在厨房能做和不能做的事。（Help Jack list something he can do and something he can't in the kitchen according to the dialogue.）

Dos：Store food in a clean and dry container before put it in the refrigerator.

Label the food container with contents and date.

Spoon from the soup into one spoon and then put the food you wanna taste into another small spoon before you taste it.

Don'ts：Never taste the soup over an open stockpot.

（此题可以让学生根据对话内容先找出关键句，再用自己的话说，表达方式不唯一）

（2）你能列出更多厨房准则吗？跟你的搭档讨论一下。（Can you list more dos and don'ts in the kitchen? Discuss with your partner.）

此题目根据基础模块中的厨师岗位英语中的厨师岗位职责部分学习内容来说即可。比如：Use clean, running water to rinse. / Wash all fruits and vegetables / Keep raw food separated from cooked food

四、厨房安全
Safety in the kitchen

练习 Exercises

（1）根据对话内容，给出小建议。（Give the tips according to the dialogue.）

此题根据对话，言之有理即可

How to avoid getting burnt in the kitchen? 如何避免在厨房里烧伤？

You have to cover as much of your skin as you can to avoid burns and if you have a special chef jacket, it must be all cotton.

How to avoid getting cut in the kitchen? 如何避免刀伤？

Keep all your fingers out away from the knife and you always have the knife in contact with your hand.

（2）根据下面给出的小建议造对话。（Make a dialogue according to the tips given below.）

（答案略）

五、项目九总练习
Exercises for part IX

Translate the following into English.

（1）Can you talk about the process of becoming a chef?

（2）You do get paid by your employer but it's a really small wage.

（3）It really helped me a lot to learn much experience.

（4）I hope I can have a chance to enjoy a bite by your cooking.

（5）I think you'll adapt quickly.

（6）That's so nice of you.

（7）It should be labeled with the contents and date.

（8）Do remember, never taste the soup over an open stockpot.

（9）Thanks for the safety tips.

（10）I'll bear in mind.

第三部分 拓展模块
Part III Extended Module

项目十 西式餐点简介
Brief Introduction of Western Cuisine

一、西餐基本烹饪法
Basic Cooking Methods of Western Cuisine

学生学习 Student Learning

导入 Lead-in

◎ 思考题 Questions

① 参照知识背景介绍部分相关内容。

② 答案不唯一，言之有理即可。

练习 Exercises

课文回顾：

西餐基本烹饪法分为湿热法和干热法两种。其中，湿热法又可以分为过水、过油、低温水煮、沸煮、蒸和焖，干热法则可以分为烤、烘烤、炙烤、炭烤、翻炒、油炸和煎。实例只要合理即可。

二、西餐食品原料
Raw Food Material of Western Cuisine

学生学习 Student Learning

导入 Lead-in

◎ 思考题 Questions

① 详见基础模块中食品原料章节。

② 答案不唯一，言之有理即可。

食品原料 Words-Learning: Raw Food Material

奶制品 milk products

> 试一试 Have a try
>
> whole milk　　　　　　　　　　（全脂牛奶）
> low-fat milk　　　　　　　　　（低脂牛奶）
> skim milk　　　　　　　　　　（撇取牛奶）
> ice milk　　　　　　　　　　　（冷冻牛奶）
> sherbet　　　　　　　　　　　（冷冻果汁牛奶）

练习 Exercises

（1）turkey 火鸡　　cheese 奶酪　　lobster 龙虾　　potato 马铃薯

（2）freshwater fishes　　　淡水鱼
　　 saltwater fishes　　　海水鱼
　　 shellfish　　　　　　贝壳水产品
　　 crustacean　　　　　甲壳水产品
　　 mollusk　　　　　　软体水产品

三、西餐烘焙简介
The Brief Introduction of Baking

学生学习 Student Learning

1. 常用设备 Common Facilities

练习 Exercises

根据图片，写出设备英文名。

piping tube; electric egg beater / mixer; spatula; scale; measuring cup; oven

2. 常用原料 Common Raw Material

练习 Exercises

（1）cocoa powder; flour; almond; shredded coconut

（2）brown sugar　　　　红糖
　　　cube sugar　　　　　方糖
　　　icing sugar　　　　　糖粉
　　　malt sugar　　　　　麦芽糖
　　　granulated sugar　　砂糖
　　　lactose　　　　　　　乳糖

3. 成品 Products

练习 Exercises

pudding; fruit pie; French bread; tart; cookie; finger sponge; donut; toast

四、西餐用餐礼仪
Dining Etiquette of Western Cuisine

学生学习 Student Learning

练习 Exercises

（1）Choosing a Restaurant　　选择餐馆
（2）Reserving a Table　　　　 预订餐位
（3）Arriving at Restaurant　　 到达餐馆
（4）Summoning a Waiter　　　召唤侍者
（5）Ordering Food　　　　　　点餐
（6）Enjoying the Meal　　　　 用餐
（7）Paying the Bill　　　　　　付款

项目十一　西餐主题口语对话
Dialogues with the Topics of Western Cuisine

一、点餐
Taking Orders

练习 Exercises

（1）How is the steak prepared?

　　rare / well-done / medium

（2）Questionnaire.

参照课文内容，根据实际情况作答。

二、菜肴制作
The Cooking Ways

1. 汤 Soup

练习 Exercises

（1）Find out the right word for each picture.

Western Cook; French onion Soup; Chinese Cook

（2）Write the recipe according to the dialogue for a Chinese cook.

①原材料：洋葱40个、黄油、牛肉汤、水、月桂叶、胡椒、百里香、法式面包、瑞士帕玛森乳酪（40 onions, butter, beef broth, water, bay leaves, pepper and thyme, French bread, Swiss and Parmesan cheese）

②制作过程：将四十个洋葱切片，然后用黄油烹调切片洋葱。翻炒20至30分钟。加入牛肉汤，水，月桂叶，胡椒和百里香。接着加热至水开。调小火，盖上盖子慢炖。在此期间，将法式面包烤约15分钟。然后把法式面包放入碗中，把洋葱汤浇在面包顶部。最后一步是在顶部撒上瑞士帕玛森乳酪，在上桌前放在烤箱内。

（Slice 40 onions, then cook the onions in butter. Stir for 20 or 30 minutes. Add the beef broth, water, bay leaves, pepper and thyme. Then heat it to boiling. Reduce the heat, cover and simmer. And in the meantime, toast the French bread about 15 minutes. Then put the toasted French bread into bowls and poured onion soup on top. The last step is to sprinkle Swiss and Parmesan cheese on top and put the bowls in the salamander just before serving.）

2. 甜点 Dessert

练习 Exercises

Material: flour, sugar, coca, baking soda, salt, baking powder, water, shortening, eggs and vanilla, whipping cream, cherry filling and frosting, and garnish with chocolate curls and Maraschino cherries.

Step 1, first preheat the oven to 350°F. And grease and flour two cake pans.

Step 2, then put the flour, sugar, coca, baking soda, salt, baking powder, water, shortening, eggs and vanilla in the mixer. Mix the ingredients at low speed for 30 seconds.

Step 3, next you should mix the ingredients at high speed for 3 minutes. Then pour them into baking pans. And put the two pans in the oven for about 30 minutes.

Step 4, in the meantime, you can prepare the cherry filling.

Step 5, after the cake is ready, put the cake on wire racks to cool.

Step 6, when it have cooled, you can spread the whipping cream, cherry filling and frosting, and garnish with chocolate curls and Maraschino cherries.

三、项目十一总练习
Exercises for part XI

（1）May I have a look at the menu first?

（2）How would you like it prepared? Rare, medium or well-done?

（3）Add the beef broth, water, bay leaves, pepper and thyme.

（4）In the meantime, you can toast the French bread.

（5）Preheat the oven to 350°F.

（6）After the cake is ready, put the cake on wire racks to cool.

四、单词表
Vocabulary

A

abalone [ˌæbəˈləuni] n. 鲍鱼
achieve [əˈtʃiːv] v. 达到，实现
actually [ˈæktʃuəli] adv. 事实上
adapt [əˈdæpt] v. 适应
add [æd] vt. & vi. 加，加入；增加，添加
advanced [ədˈvɑːnst] adv. 先进的
afterwards [ˈɑːftəwədz] adv. 后来
agar [ˈeɪɡɑː(r)] n. 琼脂
aged [ˈeɪdʒid] adj. 年老的

almond [ˈɑːmənd] n. 杏仁
American [əˈmerikən] adj. 美国的
antler [ˈæntlə] n. 茸角
apart from 除……之外
asparagus [əˈspærəɡəs] n. 芦笋；芦笋的茎
assistant [əˈsɪstənt] n. 助理
aroma [əˈrəumə] n. 芳香
attorney [əˈtɜːni] n. 律师
automatic [ˌɔːtəˈmætik] adj. 自动的
available [əˈveɪləbəl] adj. 可得到的

avoid [əˈvɔid] v. 避免
aweto [ɑːˈwetəu] n. 冬虫夏草

B

baguette [bæˈget] n. 法棍面包
bake [beik] v. 烤, 烘焙
baking soda n. 碳酸氢钠, 发酵粉
bamboo [bæmˈbuː] n. 竹, 竹竿
bandage [bændidʒ] n. 绷带
basket [ˈbɑːskit] n. 篮, 筐
bass [bæs;beis] n. 鲈鱼
battle [ˈbætl] n. 战斗, 战役; 交战 vt. & vi. 与(对)…作战, 争斗
bay [bei] n. 月桂树
bean [biːn] n. 豆, 豆科植物
beater [ˈbiːtə(r)] n. 搅拌器
beef [biːf] n. 牛肉
bell [bel] n. 钟状物
belly [ˈbeli] n. 腹部的肉
besides [biˈsaidz] adv. 而且, 还有
beverage [ˈbevəridʒ] n. 饮料
birthday [ˈbɜːθdei] n. 生日
biscuit [ˈbiskit] n. 饼干
bisque [bisk] n. 浓汤
bit [bit] n. 少量, 少许; 小片, 小块
bitter [ˈbitə] adj. 苦的
black [blæk] adj. 黑色的; n. 黑色
black bean n. 豆豉; 黑大豆
black forest cake n. 黑森林蛋糕
blade [bleid] n. 刀刃, 刀片
blanch [blɑːntʃ] v. 过水; 漂白
blood curd [blʌd] [kəːd] n. 血旺
blunt [blʌnt] adj. 钝的
boil [bɔil] n. 煮沸 vt. & vi. (使)沸腾; 开 vt. 用开水煮, 在沸水中煮
boiler [ˈbɔilə] n. 锅炉
boletus [bəuˈliːtəs] n. 牛肝菌
bottom [ˈbɒtəm] n. 底部
bowl [bəul] n. 碗, 钵, 盘; 一碗之量
brain [brein] n[c]. 脑
braise [breiz] vt. (用文火)炖; 烧
bran [bræn] n. 麸, 糠
branch [brɑːntʃ] n. 树枝, 枝条
breadcrumb [ˈbredkrʌmb] n. 面包屑
breast [brest] n[c]. 胸部

brief [briːf] adj. 短暂的; 简短的 n. 概要, 摘要
briefly [ˈbriːfli] adv. 简单地
brisket [ˈbriskit] n[c]. 胸(通常指牛肉)
broad [brɔːd] adj. 宽的, 辽阔的
broccoli [ˈbrɒkəli] n. 西蓝花, 花椰菜
broil [brɔil] vt. 烤, 炙
broth [brɔːθ: brɔθ] n. 肉汤, 鱼汤, 菜汤
brown [braun] adj. 棕色的
Buddha [ˈbudə] n. 佛陀; 佛像
buffet [ˈbʌfit] n. 自助餐
bulb [bʌlb] n. 鳞茎
bullfrog [ˈbulfrɒg] n. 牛蛙
butter [ˈbʌtə] n. 黄油; 黄油状的食品 vt. 抹黄油于…上
butterfly [ˈbʌtəflai] n. 蝴蝶
by heart 牢记

C

cabbage [ˈkæbidʒ] n. 卷心菜
cabinet [ˈkæbinit] n. 橱柜
Caesar [ˈsiːzə] n. 恺撒(罗马皇帝)
candy [ˈkændi] n. 糖果
Cantonese [ˌkæntəˈniːz] n. 广东人, 广东话
carp [kɑːp] n. 鲤鱼
carry out 执行
casserole [ˈkæsəˌrəul] n. 焙盘; 砂锅; 焙盘菜; 砂锅菜
cater [ˈkeitə] vt. & vi. 提供饮食及服务
cater to 投合, 迎合
celery [ˈseləri] n. 芹菜
cereal [ˈsiəriəl] n. 谷物食品
chafe [tʃeif] v. 摩擦
chairman n. 主席
characteristic [ˌkæriktəˈristik] n. 性格
check [tʃek] vt. & vi. 检查, 核对 n. 〈美〉支票, 账单
cheese [tʃiː(ː)z] n. 乳酪, 干酪
cheese cake 慕斯蛋糕
chef [ʃef] n. 厨师
cherry [ˈtʃeri:] n. 樱桃
chestnut [ˈtʃesnʌt] n. 栗子
chicken [ˈtʃikin] n. 鸡肉
chili [ˈtʃili] n. 红辣椒, 辣椒
Chinese yam n. 山药
chip [tʃip] v. 削; n. 炸马铃薯条

chives [tʃaivz] n. 韭黄
chocolate ['tʃɔkəlit] n. 巧克力；巧克力糖
chop [tʃɔp] vt. & vi. 砍，伐，劈
chopping block 砧板
chunk [tʃʌŋk] n. 块（较大的，不规则的）
clam [klæm] n. 蛤蚌 [c]；（供食用的）蛤肉 [u]
claw [klɔ:] n. 爪，钳
cleaver ['kli:və] n. 切肉刀
cocoa ['kəukəu] n. 可可粉
coconut milk 椰奶
colleague ['kɔli:g] n. 同事
combine [kəm'bain] v. 结合 n. 集团
commis [kə'mi:] n. 厨助；学徒工
complicated ['kɔmplikeitid] adj. 结构复杂的
condensed [kən'denst] adj. 浓缩的
congee ['kɔndʒi:] n. 粥
container [kən'teɪnə(r)] n. 容器
content [kən'tent] n. 内容
convection [kən'vekʃən] n.（热的）对流
cookie ['kuki] n. 饼干
cotton ['kɔtn] adj. 棉制的
crab [kræb] n. 螃蟹 [c]；蟹肉 [u]
crab [kræb] n. 蟹
cream [kri:m] n. 乳脂，奶油；乳霜，乳膏
creativity [kri:eɪ'tɪvəti] n. 创造力
crisp [krisp] adj. 脆的，鲜脆的
croaker ['krəukə] n. 黄花鱼
croissant [krwɑ'sɑn] n. 羊角面包；（法）新月形面包
crucian ['kru:ʃən] n. 欧洲鲫鱼
crucian carp n. 鲫鱼
crustacean [krʌ'steiʃən] 甲壳水产品
crystal ['kristəl] n. 水晶；结晶
cube [kju:b] 块（立方块）
cucumber ['kju:kʌmbə] n. 黄瓜，胡瓜
cuisine [kwɪ'zi:n] n. 烹饪艺术；菜肴
cumin ['kʌmin] n. 小茴香，孜然
curd [kə:d] n. 凝乳
curl [kə:l] n. 一绺鬈发；卷曲物，螺旋状物
curry ['kə:ri] n. 咖喱
custard ['kʌstəd] n. 奶油冻
cuttlefish ['kʌtl‚fiʃ] 墨鱼

D

daisy ['deizi] 菊科植物
Danish ['deiniʃ] adj. 丹麦的；丹麦人的；丹麦语（文）的 n. 丹麦语
date [deit] n. 日期
deep frier 油锅，深炸（油）锅
deer [diə] n. 鹿
dessert [di'zə:t] n.（餐后）甜食，甜点
detergent [di'tə:dʒənt] n. 洗涤剂
diagonal [daɪ'æɡənəl] n. <数>对角线；斜线 adj. 对角线的 adv. 对角线地
dice [dais] vt. 将…切成小方块，切成丁
diced [daist] adj. 切粒的
discount ['diskaunt] n. 数目，折扣 vt. & vi. 打折扣，减价出售
dishwasher ['diʃ‚wɔʃə] n. 洗碗碟机
disinfect [‚disin'fekt] vt. 消毒，杀菌
donut ['dəunət] n. 炸面圈
double-cooked pork slices 回锅肉
dough [dəu] n. 生面团
dragon ['drægən] n. 龙
drain [drein] vt. & vi.（使）流干，（使）逐渐流走 vt. 喝光，喝干 n. 排水沟，排水管
dried [draid] adj. 弄干了的
dryer ['draiə] n. 干燥机
duck [dʌk] n. 鸭，母鸭；鸭肉

E

eagle-eyed ['i:gl‚aid] adj. 有眼力的
ear [iə] n. [c] 耳朵
edge [edʒ] n. 边
effective [i'fektiv] adj. 有效的
eggplant ['egplɑ:nt] n. 茄子
electric [i'lektrik] adj. 电的
emphasize ['emfəsaiz] vt. 强调；加强语气；重读
employer [ɪm'plɔɪə(r)] n. 雇主
en casserole [ɔŋkɑ:s'rɔl] [法语]（菜肴）砂锅烧的，罐焖的，焙盘烤的
especially [is'peʃəli] adv. 尤其是
evaporated [i'væpəreitid] adj. 浓缩的，脱水的，蒸发干燥的
even ['i:vən] adj. 均匀的；有规律的；稳定的
executive [ig'zekjətɪv] n. 行政领导

F

fascinating ['fæsineitiŋ] adj. 迷人的，有极大吸引力的
feet ['fi:t] n. [pl] 脚

ferment [ˈfəːment] v. 发酵
fermented [fəˈmentid] adj. 发酵的
fern [fəːn] n. 蕨类植物
filet [fiˈlei] 片
fillet [ˈfilit] n. 束发带，鱼肉片
filling [ˈfiliŋ] n. 填充
filter [ˈfiltə] n. 过滤，过滤器 vt. & vi. 透过，过滤
fingernail [ˈfiŋɡənei] n. 手指甲
firewood [ˈfaiəˌwud] n. 木柴
flame [fleim] n. 火焰
flavor =flavour[ˈfleivə] n. 味；味道；特色，特性，气氛
flatfish [ˈflætˌfiʃ] n. [鱼] 比目鱼（鲽形目鱼的总称）
flounder [ˈflaundə] n. [c] 比目鱼，龙利
flour [ˈflauə] n. 面粉；粉状物质
foie gras [ˈfwaːˈɡraː] 鹅肝酱
foil [fɔil] n. 箔，金属薄片
forest [ˈfɔrist] n. 森林，丛林
frankly [ˈfræŋkliː] adv. 直率地（说），坦诚地
French [frentʃ] adj. 法国的，法国人的
French fries 炸薯条，炸薯片
fried string beans 干煸四季豆
frosting [ˈfrɔstiŋ] n. 霜状白糖，玻璃粉，无光泽面
fry [frai] v. 油炸；油煎
fungus [ˈfʌŋɡəs] n. [u / c] 菌类（pl, fungi）

G

garlic [ˈɡɑːlik] n. 大蒜
garnish [ˈɡɑːniʃ] vt. 给（上餐桌的食物）加装饰 n.（为色香味而添加的）装饰菜
gelatin [ˈdʒelətin] n. 明胶
ginger [ˈdʒindʒə] n. 生姜
gizzard [ˈɡizəd] n. [c] 胃
gluten [ˈɡluːtən] n. 面筋
glutinous [ˈɡluːtinəs] adj. 黏性的
go ahead 开始
golden [ˈɡəuldən] adj. 金色的
goose [ɡuːs] n. 鹅肉
gourd [ɡuəd] n. 葫芦
granulated [ˈɡrænʒəˌleitid] adj.（白糖）成砂状的
gravy [ˈɡreivi] n. 肉汁，肉卤，调味汁
grease [ɡriːs] n. 动物油脂；油膏，油脂 vt. 涂油脂于，用油脂润滑
greasy [ˈɡriːzi] a. 油腻的
grill [ɡril] v. 烧，烤

grinder [ˈɡraində] n. 研磨器

H

Hakka [ˈhɑːkˈkɑː] n.〈汉〉客家，客家人，客家语
ham [hæm] n. 火腿
heart [hɑːt] n. [c] 心
heal [hiːl] n. 痊愈
herring [ˈheriŋ] n. 青鱼
hilsa [ˈhilsə] n. 花点鲥属
hilt [hilt] n. 刀把
hock [hɔk] 肘关节
honey [ˈhʌni] 蜂蜜
hood [hud] n. 排风罩
hot [hɔt] adj. 热的；辣的
hotbed [ˈhɔtbed] n. 温床
hotbed chives 韭黄
hygiene [ˈhaidʒiːn] n. 卫生

I

ice breaker n. 碎冰器
ice cream n. 冰淇淋
icing [ˈaisiŋ] 装饰品；锦上添花
icing sugar n. 高度精炼的磨的极细的粉末状蔗糖（糖霜）
immediately [iˈmiːdiətli] adv. 立刻地
impatience [imˈpeiʃəns] n. 不耐烦、急躁
impressive [imˈpresiv] adj. 令人印象深刻的
in the meantime 同时
including prep. 包括；包含
inflammable [inˈflæməbl] adj. 易燃的
ingredient [inˈɡriːdjənt] n.（混合物的）组成部分；配料
instant [ˈinstənt] adj. 立即的，即刻的
internship [ˈintəːnʃip] n. 实习
intestine [inˈtestin] n. [c] 肠
introduce [ˌintrəˈdjuːs] v. 介绍
introduction [ˌintrəˈdʌkʃən] n. 介绍，引见
iron [ˈaiən] adj. 铁的

J

jar [dʒɑː] n. 盅，罐子，广口瓶；（啤酒）杯
jellied [ˈdʒelid] adj. 凝成胶状的
Jellyfish n. 海蜇
juice [dʒuː(ː)s] n. 果汁
jumbo [ˈdʒʌmbəu] adj. 巨大的；特大的

K

kelp [kelp] *n.* 海带
kid [kid] *n.* 小孩；年轻人
kidney ['kidni] *n.* 肾脏
kitchen ['kitʃin] *n.* 厨房
knife [naif] *n.* 刀
Kung Pao Chicken 宫保鸡丁

L

label ['leibl] *v.* 把…贴标签
lactose ['læktəus] *n.* 乳糖
lamb [læm] *n.* 羊羔肉
larder ['lɑ:də(r)] *n.* 肉贮藏处
lean [li:n] *adj.* 瘦的
leg [leg] *n.* [u / c]（猪、羊等）供食用的腿
lend [lend] *vt.* 增加，增添
level ['levl] *n.* 水平
light soy sauce 生抽
lily ['lili] *n.* 百合花
lime [laim] *n.* 石灰；酸橙；绿黄色
lip [lip] *n.* 嘴唇
liquor ['likə] *n.* 酒，含酒精饮料
liver ['livə] *n.*（供食用的）肝
lobster ['lɔbstə] *n.* 龙虾 [c]；龙虾肉 [u]
lotus ['ləutəs] *n.* 莲花，荷花
lung [lʌŋ] *n.* [c] 肺
lure [ljuə] *vt.* 诱惑；引诱

M

macaroni [ˌmækə'rəuni] *n.* 通心粉；通心面条
machine [mə'ʃi(:)n] *n.* 机器
mackerel ['mækrəl] *n.* 马鲛鱼（单复数同形）
Maggi [mædʒi] *n.* 美极（厨房调理食品品牌名）
mainly ['meinli:] *adv.* 大部分地；主要地，在多数情况下
maintain [mein'tein] *vt.* 保持
malt [mɔ:lt] *n.* 麦芽
mandarin ['mændərin] *n.*（中国的）官话（普通话的旧称）
mandarin fish *n.* 鳜鱼
mango ['mæŋgəu] *n.* [园艺] 芒果
Mapo Tofu *n.* 麻婆豆腐
Maraschino cherries 马拉斯加樱桃，酒渍糖水樱桃
maraschino [ˌmærə'ski:nəu] *n.* 黑樱桃酒
margarine [ˌmɑ:dʒə'ri:n] *n.* 人造黄油
marinate ['mærineit] *v.* 浸泡在卤汁中在鱼或肉上浇上卤汁
mashed [mæʃt] *adj.* 捣碎的；捣烂的
maw [mɔ:] *n.*（动物的）胃
meantime ['mi:ntaim] *adv.* 同时；其间 *n.* 其时，其间
measure ['meʒə(r)] *v.* 测量
medium ['mi:djəm] *adj.* 中等的，适中的
melon ['melən] *n.* 瓜，甜瓜
menu ['menju:] *n.* 菜单；（荧光屏上显示的）项目
meter ['mi:tə] *n.*〈美〉米；计，表，仪表
Mexican ['meksikən] *adj.* 墨西哥的；墨西哥人的 *n.* 墨西哥人；墨西哥语
microwave ['maikrəweiv] *n.* 微波
millet ['milit] *n.* 小米
mince [mins] *vt.* 切碎，剁碎，绞碎 *n.* 肉末
minced [minst] *adj.* 切碎的；切成末的
minestrone [ˌmini'strəuni] *n.* 蔬菜通心粉汤（意大利美食）；蔬菜面条汤
mixer ['miksə] *n.* 搅拌器，混合器
mollusc ['mɔləsk] *n.* 软体水产品
moss [mɔs] *n.* 苔藓
mousse [mu:s] *n.* 慕斯；奶油冻
mung [mʌŋ] *n.* 绿豆
mushroom ['mʌʃrum] *n.* 蘑菇
mustard ['mʌstəd] *n.* 芥末
mustard greens with scallop 干贝芥菜（干贝）
mutton ['mʌtən] *n.* 羊肉

N

napkin [næpkin] *n.* 餐巾
necessary ['nesəˌsəri] *adj.* 必要的；必然的 *n.* [pl.] 必需品
nest [nest] *n.*（鸟）窝，巢
new-comer *n.* 新来的人
New Zealand *n.* 新西兰
northern ['nɔ:ðən] *adj.* 北方的，北部的
nut [nʌt] *n.* 坚果

O

oatmeal ['əutmi:l] *n.* 燕麦片
occasional [ə'keizənəl] *adj.* 偶尔的，不经常的
occupy ['ɔkjupai] *vt.* 占领；占有；使从事；使忙于
oil [ɔil] *n.* 油；石油；油画；油画颜料
oil smoker exhausting equipment *n.* 抽排油烟设备
on the rocks 加冰块

onion [ˈʌnjən] n. 洋葱（头）；（食物）洋葱，葱头
osmanthus [ɔzˈmænθəs] n. 桂花
oval [ˈəuvəl] adj. 椭圆的
oven [ˈʌvən] n. 烤箱，烤炉
overcome [ˌəuvəˈkʌm] v. 克服
oyster [ˈɔistə] n. 蚝；牡蛎

P

pan [pæn] n. 平底锅、盘子
Parmesan [ˌpɑːmiˈzæn] n. 用脱脂乳制成的坚硬的意大利干酪
paste [peist] n. 膏；糊状物
pastry [ˈpeistri] n. [u / c] 油酥点心
pastry [ˈpeistri] n. 糕点
patient [ˈpeʃənt] adj. 有耐心的
peanut [ˈpiːnʌt] n. 花生；落花生
pear [pɛə] n. 梨
pepper [ˈpepə] n. 胡椒粉；辣椒 vt. 在…上撒胡椒粉；使布满
perfect [ˈpəːfikt] adj. 完美的，完善的，理想的
phoenix [ˈfiːniks] n. 凤凰
pickle [ˈpikl] n. 腌汁，泡菜；v. 腌，泡
pickled [ˈpikld] adj. 腌制的
pie [pai] n. 馅饼，派
pig [pig] n. 猪；猪肉
pigeon [ˈpidʒin] n. 鸽子
pineapple [ˈpainˌæpl] n. 菠萝
pineapple pie n. 菠萝派
piping [ˈpaipiŋ] v. 用管道输送
platter [ˈplætə] n. 大浅盘
pleasure [ˈpleʒə(r)] n. 快乐
plum [plʌm] n. 李子；梅子
poach [pəutʃ] v. 水煮
poach [pəutʃ] v. 在这里指"用热油烹饪"
point [pɔint] n. 尖，顶
pomfret [ˈpɔmfrit] n. 鲳鱼
popcorn [ˈpɔpkɔːn] n. 爆米花
pork [pɔːk] n. 猪肉
porter [ˈpɔːtə(r)] n. 搬运工
position [pəˈziʃən] n. 方位，位置；地位，身份
pot [pɔt] n. 壶；盆；罐
potato [pəˈteitəu] n. 马铃薯
pottery [ˈpɔtəri] n. 陶器
poultry [ˈpəultri] n. 家禽，禽肉
pour [pɔː] vt. & vi. 涌出；倾；倒

power [ˈpauə(r)] n. 电力
powder [ˈpaudə(r)] n. 粉
prawn [prɔːn] n. [c] 明虾，对虾
prefer [priˈfəː] vt. 选择某事物（而不选择他事物）；更喜欢
preheat [ˌpriːˈhiːt] vt. 预热
prepare [priˈpɛə] vt. 准备 vt. & vi. 筹备，进行各项准备工作
preserve [priˈzəːv] v. 腌制
preserved [priˈzəːvd] adj. 腌制的
press [pres] v. 压平
price [prais] n. 价格，价钱；代价
procedure [prəˈsiːdʒə(r)] n. 程序
process [ˈprəuses] n. 过程
pudding [ˈpudiŋ] n. 布丁
puff [pʌf] n. 蓬松；泡芙

Q

quail [kweil] n. 鹌鹑
qualified [ˈkwɔlifaid] adj. 有资格的，能胜任的
Quiche Lorraine 洛林乳蛋饼；法式蛋塔；洛林糕（用干酪和腌肉等做成的奶蛋糕）
quick [kwik] adj. 快的，迅速的

R

rabbit [ˈræbit] n. [c] 兔子
rabbit 兔肉
radish [ˈrædiʃ] n. [c] 萝卜
rare [rɛə] adj. 稀少的，罕见的；稀薄的
raw [rɔː] adj. 生的
realize [ˈriəlaiz] v. 意识到
recommend [ˌrekəˈmend] vt. 推荐，介绍
red bean n. 红豆
reduce [riˈdjuːs] vt. 缩减，减少；降低
refrigerator [riˈfridʒəreitə] n. 冰箱
relief [riˈliːf] n. 换班者
relish [ˈreliʃ] n. 调味品
requirement [riˈkwaiəmənt] n. 要求，必要条件
reserve [riˈzəːv] vt. 保留；预订
reserved [riˈzəːvd] adj. 预订的，保留的
restaurant [ˈrestərənt] n. 饭店，餐馆
rib [rib] n. [c] 排骨；肋骨
ribbonfish [ˈribənfiʃ] n. 带鱼
rice wine n. 米酒
rim [rim] n. 边，边缘 [c]

rinse [rɪns] vt. 漂洗，冲洗；用清水漂洗掉（肥皂泡等）
roast [rəust] vt. & vi. 烤；烘；焙 adj. 烤好的，烤制的
roast suckling pig 烤乳猪
rock [rɔk] n. 岩石
roe[rəu] n. 鱼子
roll[rəul] n. 卷
root [ru:t] n. 根
rye [raɪ] n. 黑麦

S

saffron [ˈsæfrən] n. 藏红花
safety [ˈseɪfti] n. 安全
sago [ˈseigəu] n. 西米，西谷米
sake [seik] n. 日本清酒
salad [ˈsæləd] n. 沙拉，凉拌菜
salamander [ˈsæləˌmændə] n. 制作糕饼的烤箱
salt [sɔ:lt] n. 盐
salt-baked chicken 盐焗鸡
salty [ˈsɔ:lti] adj. 含盐的，咸的
satisfaction [ˌsætisˈfækʃən] n. 满意，满足，实现；赔偿物，补偿
sauce [sɔ:s] n. 酱油；少司；调味汁
saucepan [ˈsɔ:spæn] n. 深平底锅
sauté [ˈsəutei] v. 炒；嫩煎
scald [skɔ:ld] v. 煮沸；烫
scald [skɔ:ld] vt. （沸水等）烫伤（皮肤）；把（尤指牛奶）加热到接近沸腾
scale [skeɪl] n. 刻度；天平
scallion [ˈskæljən] n. 青葱，冬葱，大葱
scallop [ˈskɔləp] n. 扇贝壳 v. 拾扇贝
scoop [sku:p] n. 铲子；勺
scorpion [ˈskɔ:pjən] 蝎子 [c]
scramble [ˈskræmbl] vt. 扰乱，搞乱 vi. /n. 攀登，爬行；争夺
seafood [ˈsi:ˌfu:d] n. 海产食品，海鲜
seasoned [ˈsi:zənd] adj. 经验丰富的；老练的；调过味的
seasoning [ˈsi:zənɪŋ] n. 调味品，佐料
seating [ˈsi:tɪŋ] n. 座位；席位
seed [si:d] n. 种子
self-introduction[selfˌintrəˈdʌkʃən] n. 自我介绍
serve [sə:v] vt. & vi. （为…）服务；任（职）；提供 vt. 向…供应
sesame[ˈsesəmi] n. 〈植〉芝麻，脂麻

shank [ʃæŋk] n. 小腿
sharp [ʃɑ:p] adj. 锋利的
shellfish [ˈʃelfiʃ] n. 贝，甲壳类动物
sherbet [ˈʃə:bət] n. 冰果子露，冷冻牛奶
shoot [ʃu:t] n. 嫩芽
shred [ʃred] n. 碎片，细条，破布；些许，少量 vt. & vi. 撕碎，切碎
shredded [ʃredid] adj. 切丝的；切碎的
shrimp [ʃrimp] n. [c] 基围虾
similar [ˈsimilə] adj. 类似的；同类的；相似的；同样的
simmer [ˈsimə] n. 炖 vt. & vi. 炖；慢煮
situation [ˌsitjuˈeiʃən] n. 情况
sirloin steak 沙朗牛排，菲力牛排
skill [skɪl] n. 技能；本领
skimmer [ˈskimə] n. 撇沫器
skin [skin] n. 皮（肤）；兽皮，毛皮；（蔬菜，水果等）外皮，外壳
slice [slais] vt. 切 n. 片，薄片，切片；部分；份
sliced [slaist] adj. 切成薄片的
smoke [sməuk] v. 熏
snow [snəu] n. 雪
soaked [səukt] adj. 湿透的，浸透的 v. 浸湿（soak 的过去分词）
soda [ˈsəudə] n. 苏打，碱；苏打水，汽水
soup [su:p] n. 汤，羹
sour [ˈsauə] adj. 有酸味的，酸的
sous [su:] n. 担任助理的
soy [sɔi] n. 大豆，豆酱
soy sauce n. 酱油
spacious [ˈspeiʃəs] adj. 宽敞的
spaghetti [spəˈgeti] n. 意大利式细面条
sparrow [ˈspærəu] n. [c] 麻雀
spatula [ˈspætʃələ] n. 抹刀
spicy [ˈspaisi] adj. 辛辣的；香的，多香料的
spider [ˈspaidə] n. 蜘蛛
spine [spain] n. [c]（动、植物的）刺，针
spirit [ˈspirit] n. 酒精
split [split] vt. & vi. （使）裂开；（使）破裂
spoil [spɔil] vt. 损坏；毁掉
sponge [spʌndʒ] n. 松软布丁
spread [spred] vt. & vi. 伸开，展开，摊开；（使）传播，（使）散布
sprinkle [ˈsprɪŋkl] vt. & vi. 洒，撒
sprout [spraut] n. 芽，苗芽

square [skwɛə] adj. 方的
squid [skwid] n. 鱿鱼
squirrel ['skwirəl] n. 松鼠
staff [stɑ:f] n. 职员
starch [stɑ:tʃ] n. 淀粉
steady ['stedi] adj. 稳的，稳定的，坚定的；不变的
steak [steik] n. 牛排
steam [sti:m] vt. 蒸煮
steamer ['sti:mə] n. 汽锅，蒸锅
stew [stju:] vt. & vi. 炖，焖
steward ['stju:əd] n. 膳务员
stick [stik] n. 棍
sticky ['stiki] adj. 黏性的
stir [stə:] vt. & vi. 搅拌
store [stɔ:(r)] v. 储藏
stone [stəun] adj. 石的，石制的
stove [stəuv] n. 炉子
straight up 净饮
straw [strɔ:] n. 稻草
strawberry ['strɔ:bəri] n. 草莓
strict [strikt] adj. 严格的
string bean 四季豆
strongpoint ['strɔŋˌpoint] n. 优点
successful [sək'sesfəl] adj. 成功的
suckle ['sʌkl] vt. & vi. 给…喂奶
suckling pig 乳猪
sugar ['ʃugə] n. 食糖
summon ['sʌmən] vt. 召唤，传唤
surface ['sə:fis] n. 面，表面；水面，液体的表面；外表，外观
swallow ['swɔləu] n. 燕子
swallow's nest porridge 燕窝粥
sweet [swi:t] adj. 甜的
Swiss [swis] adj. （人）来自瑞士的；瑞士的；瑞士文化的
syrup ['sirəp] n. 糖浆

T

tablespoon ['teibəlˌspu:n] n. 大汤匙，大调羹；一大汤匙的量
tableware ['teiblwɛə] n. 餐具
tail [teil] n. [c] 尾巴
tart [tɑ:t] 果馅饼
taste [teist] n. 滋味，味道；鉴赏力；爱好，嗜好
tea [ti:] tree [tri:] 茶树

teaspoon ['ti:ˌspu:n] n. 茶匙；一茶匙的量
technical ['teknikəl] adj. 技术的，专业的
temperature ['tempəritʃə] n. 温度，气温
tender ['tendə] adj. 嫩的
tenderloin ['tendəloin] 腰部嫩肉；里脊肉
tendon ['tendən] n. 筋腱
terrible ['terəbl] adj. 很糟糕的；可怕的
Thai [tai] adj. 泰国的
thaw [θɔ:] vi. 溶化，溶解；（气候）解冻 vt. 使融化，使缓和
thermometer [θə'mɒmitə(r)] n. 温度计
thyme [taim] n. （用以调味的）百里香（草）
timer ['taimə(r)] n. 计时器
tin [tin] adj. 锡制的
tip [tip] n. 小建议
Tiramisu [tirəmə'su:] n. 提拉米苏
tiramisu [ti'rɑ:misju:] n. 意大利式甜点
toast ['təust] n. 烤面包；吐司
tofu ['təufu:] n. 豆腐
tong [tɒŋ] n. 钳子
tongue [tʌŋ] n. [u / c] （牛等烹饪用）舌肉
toon [tu:n] 红椿木
toss [tɔ:s] v. 拌
traditional [trə'diʃənəl] adj. 传统的
transfer [træns'fə:] vt. & vi. 转移；迁移
tray [trei] n. 托盘
tripe [traip] n. 肚
tube [tju:b] n. 管状物
truffle ['trʌfl] n. [u / c] 松露
tuna ['tju:nə] n. 金枪鱼，鲔鱼
turbot ['tə:bət] n. 多宝鱼（单复数同形）
turkey ['tə:ki] n. 火鸡，火鸡肉
turtle ['tə:tl] n. 甲鱼
twisted ['twistid] adj. 扭曲的
typical ['tipikəl] adj. 典型的，有代表性的；特有的，独特的

U

uniform ['ju:nifɔ:m] n. 制服
usage ['ju:sidʒ] n. 用法

V

vanilla [və'nilə] n. 香子兰；香草；香草精 adj. 香草的；香草味的
vanilla ice cream 香草冰淇淋

variety [vəˈraiəti] n. 品种，种类；种种，各种；变化，多样化
veal [vi:l] n. 小牛肉
vegetable [ˈvedʒitəbl] n. 蔬菜
vegetarian [ˌvedʒiˈtɛəriən] n. 素食者；食草动物 adj. 素食的
vegetarian food 素食；斋饭
venison [ˈvenizən] n. 鹿肉
vinegar [ˈvinigə] n. 食醋

W

wage [weidʒ] n. 工资
walk-in [ˈwɔːkˌin] n. 可供人走进之物
wasabi [waːˈsaːbi] n. 日本芥末
water [ˈwɔːtə] n. 水
weakness [ˈwiːknis] n. 弱点
wheat [wiːt] n. 小麦
whelk [welk] n. 螺

Whisky [ˈwiski] n. 威士忌酒
white [wait] adj. 白色的
wild [waild] adj. 野生的
wine [wain] n. 酒
wing [wiŋ] n. [c] 翅膀
wire [ˈwaiə] n. 金属丝，金属线；电线，导线；电报
wolfberry [ˈwulfbəri] n.（美）西方雪果
wonderful [ˈwʌndəfəl] adj. 极好的，惊人的

Y

yam [jæm] n. 甘薯
yeast [jiːst] n. 酵母
yogurt [ˈjəugəːt] n. 酸奶，酵母乳
yolk [jəulk] n. 卵黄
Yu-Shiang Shredded Pork 鱼香肉丝

参考文献 Reference

［1］Carol Rueckert．餐厅英语情景口语50主题[M]．北京：外文出版社，2009．
［2］陈明瞭．烘焙专业英语[M]．广州：暨南大学出版社，2014．
［3］郭亚东．西餐烹调技术[M]．北京．中国轻工业出版社，2007．
［4］黄海翔．中餐菜单英译浅谈[J]．中国科技翻译，1999，12（1）：18-21．
［5］韩枫．烹调技术[M]．北京：中国劳动和社会保障出版社，2001．
［6］姜玲．厨师岗位英语[M]．北京：旅游教育出版社，2008．
［7］李柏红，张小玲．烹饪专业英语[M]．北京：中国商业出版社，2006．
［8］刘萍．中式菜肴名称的口译[J]．中国科技翻译，2003，16（3），18-20．
［9］刘清波．中式菜名英译的技巧和原则[J]．中国科技翻译，2003，16（4），52-54．
［10］李荣耀，洪锦怡，曾淑凤．西餐烹饪务实[M]．天津：南开大学出版社，2005．
［11］钱立春．烹饪原料知识[M]．北京：中国劳动和社会保障出版社，2001．
［12］乔平．中餐菜名分类及其英文方法[J]．扬州大学烹饪学报，2004（2），46-49．
［13］任静生．也谈中菜与主食的英译问题[J]．中国翻译，2001，22（6），56-57．
［14］孙诚．烹饪英语[M]．北京：高等教育出版社，2009．
［15］王天佑．西餐概论[M]．北京：旅游教育出版社，2005．
［16］夏宏伟．浅析中菜英译[J]．湖北成人教育学院学报，2009，15（5），82-84．
［17］杨柳．现代厨房管理[M]．北京：高等教育出版社，2004．
［18］周宏．烹饪原料知识[M]．北京：中国劳动和社会保障出版社，2007．
［19］赵丽．烹饪英语[M]．北京：北京大学出版社，2010．
［20］周妙林．中餐烹调技术[M]．北京：高等教育出版社，2002．
［21］张艳红．西餐英语[M]．北京：中国人民大学出版社，2012．
［22］周晓燕．烹调工艺学[M]．北京：中国轻工业出版社，2000．